MOUNTAINS ON THE MOON

Printed and bound by Antony Rowe Ltd, Chippenham, England

Published by Crossbridge Books
345 Old Birmingham Road
Bromsgrove B60 1NX
Tel: 0121 447 7897

© Michael Arthern 2004

First published 2004

All rights reserved. No part of this publication
may be reproduced, stored in a retrieval system,
or transmitted in any form or by any means – electronic,
mechanical, photocopying, recording or otherwise –
without prior permission of the publisher.

ISBN 0 9543573 7 X

British Library Cataloguing in Publication Data.
A catalogue record for this book is available from the British Library.

Also published by Crossbridge Books:

> *The God of Miracles* Trevor and Anne Dearing
> *Schizophrenia Defeated* James Stacey
> *Total Healing* Trevor Dearing (under the imprint ***Mohr Books***)
> *It's True!* Trevor Dearing (***Mohr Books***)
> *Called to Be a Wife* Anne Dearing
> *Ten For Today* David Morgan
> *The Reluctant Bride* Kamala Sabaratnam

For children aged 8-12:

> *The Kentle-Shaddy* Eileen Mohr
> *The Kentle-Shaddy Knows Something* E.M. Mohr
> *The Kentle-Shaddy in Felmunia* E.M. Mohr
> *Liam's Secret Codes* Patricia Blackledge

We express our thanks to NASA/JPL/Caltech for the cover photo.

MOUNTAINS ON THE MOON
Michael Arthern

Sketches of the scientists by Rolf Mohr

CROSSBRIDGE BOOKS

Contents

	Page
Thanks	vii
Introduction	x
An Invitation	xiv
1. **IN THE BEGINNING**	1
2. THALES c.625 - c.547 BC: Engrossed in the stars	4
3. PLATO 427 - 348 BC AND ARISTOTLE 384 - 322 BC: Every realm of nature is marvellous	8
4. THEOPHRASTUS c.371 - c.287 BC: Enquiry into plants	13
5. ARISTARCHUS c310 - c230 BC: Two great lights	17
6. ARCHIMEDES c.287 - 212 BC: The proof that will carry conviction	21
7. PTOLEMY AD c.100 - c.175: Tireless diligence in calculating	25
8. **ONE DAY IN SEVEN**	29
9. COPERNICUS 1473 – 1543: A sun-centred universe	33
10. RAUWOLF 1535 – 1596: Plants which caught his eye	39
11. GILBERT 1544 – 1603: The secrets of magnetism	.43
12. GALILEO 1564 – 1642: Four previously unknown moons	47
13. KEPLER 1571 – 1630: God is being celebrated in astronomy	53
14. HARVEY 1578 – 1657: Anatomy ... from the fabric of nature	57
15. **TWO BOOKS**	62
16. BOYLE 1626 – 1691: Touching the spring of the air	66
17. RAY 1627 – 1705: Filled with wonder and delight	70
18. NEWTON 1642 – 1727: An astonishing genius	75
19. FLAMSTEED 1646 – 1719: Looking stars in the face	79

Contents *(contd)*

20. LINNAEUS 1707 – 1778: The naming of plants and animals — 83
21. PRIESTLEY 1733 – 1804: Researches into different "airs" — 88
22. **THE BOOK OF GOD'S WORD** — 92
23. LAVOISIER 1743 – 1794: The true nature of burning — 95
24. DALTON 1766 – 1844: — .99
    Triangles, chymical processes and electrical experiments
25. SEDGWICK 1785 – 1873: — 103
    Armed only with a hammer and notebook
26. FARADAY 1791 – 1867: A good experiment would make — 109
    him almost dance with delight
27. LYELL 1797 – 1875: The greatest geologist in Europe? — 116
28. DARWIN 1809 – 1882: What a fellow for asking questions — 121
29. **THE BOOK OF GOD'S WORKS** — 129
30. JOULE 1818 – 1889: Chipping away at the confusion — 133
    surrounding heat, work and energy
31. MENDEL 1822 – 1884: Genetics in the garden — 138
32. PASTEUR 1822 – 1895: — 142
    I shall speak to you of nothing but crystals
33. KELVIN 1824 – 1907: Created for science — 146
34. MAXWELL 1831 – 1879: — 150
    Among the greatest of all intellectual achievements
35. RÖNTGEN 1845 – 1923: Electrified by his own enthusiasm — 154
36. **THE SEARCH FOR TRUTH** — 160
37. PLANCK 1858 – 1947: — 165
    Engraved on the heart of theoretical physics
38. CURIE 1867 – 1934: Science has great beauty — 169
39. RUTHERFORD 1871 – 1937: Some very big ideas — 174
40. MEITNER 1878 – 1968: — 179
    The excitement and delight of a truly new idea
41. EINSTEIN 1879 – 1955: — 185
    His search for truth was uncompromising

Contents *(contd)*

42. WEGENER 1880 – 1930: Drifting continents	190
43. **INTERPRETING SCRIPTURE**	195
44. HUBBLE 1889 – 1953: An expanding universe	200
45. LONSDALE 1903 – 1971: Excited and thrilled by new facts about crystals	206
46. CRICK 1916 – 2004 AND WATSON b. 1928: DNA: It's so beautiful, you see, so beautiful	211
47. FEYNMAN 1918 – 1988: Magic, mathematical ingenuity ... physical insight	216
48. PRANCE b. 1937: Remotest areas of the Amazon rainforest	221
49. COLLINS b. 1950: Trying to discern truth	226
50. **A NEW CREATION**	231

Biblical quotations

unless otherwise indicated are taken from the
HOLY BIBLE, NEW INTERNATIONAL VERSION
Copyright © 1973, 1978, 1984 by International Bible Society.
Used by permission.

Thanks

The birth of this book has involved many midwives and I thank them all.

Regents Park College, Oxford, granted me a studentship at the outset of the project. This gave me a base in which to work and an introduction to the Bodleian Library. I enjoyed the benefit of stimulating conversations in the Senior Common Room. I am also indebted to the College for tutorial encouragement and help from the Rev Dr John Weaver, now Principal of South Wales Baptist College.

The Radcliffe Science Library became almost a second home to me during the next few years and I found the staff helpful beyond the call of duty, among them Andrew Colquhoun in particular.

I owe a debt of kindness to several members of Southcourt Baptist Church, Aylesbury. Bernard Harlock helped me with the unfamiliar territory of proof reading and house style – Martin Manser and Steve Briars with the equally foreign world of publishing. Other friends in the church read early drafts and I am happy to thank especially Liz Stoddard and Tony Ecclestone for their encouragement and detailed constructive comments. My niece, Jenny Arthern, was also particularly helpful in this way.

As the book neared completion I sent unsolicited copies to Professor Russell Stannard and the Rev Dr David Wilkinson. Both kindly gave time to reply and their positive comments encouraged me to begin the search for a publisher.

The search led me to Eileen Mohr of Crossbridge Books and

I am more than grateful to her for ensuring that the involvement of all those I have named has not been in vain.

Because I wanted the book to end on a contemporary note after travelling through the centuries, I must especially thank Francis Collins and Ghillean Prance for their willingness to be included.

Lastly, I give thanks to and for Janet, my wife. How I came to have such a delightful helpmate through these 39 years is something I do not understand, but for which I am eternally thankful. The book is the better for changes she suggested.

It will be taken as a kindness if readers inform us of any errors they discover.

M. A.

AH, SOVEREIGN LORD,
YOU HAVE MADE THE HEAVENS AND THE EARTH
BY YOUR GREAT POWER
AND OUTSTRETCHED ARM.

Introduction

What a world we live in! It excites all our senses with its varied beauty: stars on a clear night; glittering hoarfrost; scents of honeysuckle, or of rain-drenched earth after drought; the call of the curlew; the wind in the willows; a loving touch, a pebble's smoothness; strawberries and cream, a cup of tea. Touch and taste, hearing, sight and smell combine to make us aware of its extraordinary diversity. It also prods us into astonishment, for science has revealed a vastness on the one hand and an intricacy on the other that amaze us in equal measure. The Hubble telescope and the electron microscope give us images which send our minds soaring and probing far beyond all that is available to our unaided senses.

Who are the people who have made such things possible? Who *was* Hubble? Who are the others to whom we owe our present understanding of the world? What did these people, the men and women of science, think about the world they were investigating? Did they believe in God, or is a life in science the complete turn-off from faith that some would have us accept? These questions set me on a lengthy, but intriguing journey, which has led to this book. It is not a book about the history or content of science, although some of the main strands in both will emerge as you read. It is about some of the people I met on my journey, and the thoughts they prompted, interspersed with more general comments on science and faith.

We all owe a huge debt to science; not only for the countless utilities we daily take for granted but also for the mind-blowing wonders it has unveiled. This book explores that unveiling through the lives of some of the people involved in the story of science. We owe to these people, and many more like

INTRODUCTION

them, any detailed knowledge we have of the universe; but we may know very little about them, perhaps not even their names. I have been fascinated to discover how many of them believed themselves to be investigating a created world; even, as Kepler expressed it, to be thinking God's thoughts after him. Some of them wrote from their experience as heart and soul, mind and strength Christians. Others gave a nod to 'The Creator', as the convention of the day demanded. Some were of other faiths or none. All were driven by curiosity. As a child, James Clerk Maxwell was always asking, "What's the go of that?" "What does it do?" "Yes, but what is the *particular* go of it?" [1]

Too many of us outgrow our childlike curiosity, forgetting that at the root of every discovery is a mind that has never grown tired of asking questions. The search for answers is what we now call science and may take years of reading, thinking and patient experiment, sometimes helped by a sudden flash of inspiration. Science is always building on the past, adding to our knowledge, and changing our understanding of the universe. Quite often it delivers new benefits and further challenges to morality in the same package.

The Bible, and even Christianity as a whole, are sometimes dismissed on the basis that, "Science has disproved all that". Some of the lives portrayed here expose the weakness of such statements.

Many Christians will share my belief that science and their faith can enhance each other. I hope such readers will find much here to stimulate them. Others have mixed feelings toward science. Their relationship with it has a troubled history, like that between the French and the British. This last is perhaps not so much a love-hate relationship as one of love-suspicion and for many, the relationship between science and their faith is similar. They love the benefits of science and

INTRODUCTION

marvel at its more newsworthy discoveries, but are suspicious about where it may lead and what it might do to their faith.

For some Christians, the relationship really is nearer to love-hate and the conflict may be bitter. They may enjoy the fruit of science, but hate it for its "Godless theories" or its "undermining of faith". As one such said, "I am sick and tired of letting science determine how the Bible ought to be interpreted."[2]

Some people consider that the faith-science relationship should not, or even does not, exist. Lewis Wolpert, in his fascinating book, *The Unnatural Nature of Science,* states bluntly, "Religious belief is incompatible with science", although he acknowledges the "paradox" that many scientists are deeply religious.[3] If Wolpert is wrong, there is no paradox.

Others don't care. They will make the most of whatever goodies science has on offer, but when it comes to exploring the science itself and the people behind it, there is a feeling that life is too short. Most of us probably fit somewhere in this category; but what a lot we miss.

Science itself is neutral. Different interpretations of its discoveries have nourished faith or challenged it; inspired it or destroyed it. Some have argued that Christianity has been the cradle of science in the western world. Others have written of 'warfare' and 'conflict' between theology and science.

This was particularly true as the nineteenth century drew to a close, but more than a hundred years later the legacy of old battles lingers on. Even today science can provoke great hostility, and not only from some Christians; while a few scientists are anti-Christian, even to the point where one wonders what has happened to the open mind which lies at the heart of scientific investigation. The poet Irina Ratushinskaya, as a child in the former Soviet Union, was given compulsory atheist instruction. Concerning the atheistic stance that there is no God, she wrote, "Why do grown-ups spend so much time talking about something that doesn't exist? You can't help feeling suspicious."[4]

Were I a non-Christian, or of another faith or nation, this would be, in some respects, a different book; but it would hardly be possible to write of science at all without mentioning several of those who feature here. If you find science threatening, I hope that as you explore these lives, you will come to appreciate it for what it is, an eye into a world even more wonderful than we could ever have imagined. And, Christian or not, you will find, if you read on, that many scientists have found no conflict between science and faith.

Notes

(1) J. J. Thomson in *Essays on James Clerk Maxwell. A Commemorative Volume 1831 - 1931* Cambridge University Press 1931 p 2
(2) Quoted in: D. Young *The Biblical Flood* Paternoster 1995 Preface
(3) L. Wolpert *The Unnatural Nature of Science* Faber and Faber 1992 Introduction p xiii
(4) I. Ratushinskaya *In the Beginning* Hodder and Stoughton 1990 p 26.

AN INVITATION

An invitation

Before turning to the next page, you are invited to look closely at something easily to hand; any flower or leaf, for example, which may be in the room. Perhaps there is a fish in an aquarium, or a cat asleep on a chair? Can you see a tree through the window? Better still, go outside and look up at the sky; watch the clouds, or the moon. Pick up a snail, or a pebble; a flower or a fallen leaf.

    Whatever one thing you choose, take a while to look at it. *Really* look at it. Pretend you have never seen it before. Let it raise questions in your mind. You might like to write down any questions or thoughts which occur to you. Give yourself time ...

AN INVITATION

Now read on ...

The passages below come from people who, taking time to look, gathered "the harvest of a quiet eye".

" ... most people simply don't know how beautiful the world is and how much splendour is revealed in the smallest things, in a common flower, in a stone, in the bark of a tree or the leaf of a birch. Grown-up people who have occupations and cares and who worry themselves about mere trifles, gradually lose the eye for these riches, ..."[1]

Rainer Maria Rilke
(German poet 1875-1926)

"The green trees when I saw them first ... transported and ravished me; their sweetness and unusual beauty made my heart to leap, and almost mad with ecstasy, they were such strange and wonderful things."[2]

Thomas Traherne
(English poet 1637-74)

"If we had never before looked upon the earth, but suddenly came to it man or woman grown, set down in a summer meadow, would it not seem to us a radiant vision? The hues, the shapes, the song and life of birds, above all the sunlight, the breath of heaven, resting on it; the mind would be filled with its glory, unable to grasp it, hardly believing that such things could be mere matter and no more."[3]

Richard Jefferies
(English writer/naturalist 1848-1887)

It was recounted of Brother Lawrence that "God had done him a singular favour in his conversion at the age of eighteen ... that in the winter, seeing a tree stripped of its leaves, and considering that within a little time the leaves would be renewed, and after that the flowers and fruit appear, he received a high view of the power and providence of God, which has never since been effaced from his soul. That this view had perfectly set him loose from the world, and kindled in him such a love for God that he could not tell whether it had increased during the more than forty years he had lived since."[4]

Such people, who do not "lose the eye for these riches" are sometimes, through a poem or a painting, for example, able to refocus our own dulled vision. But there are others, equally open-

AN INVITATION

eyed, whose response leaps beyond wonder to curiosity. How big is the universe? How does the moon stay in the sky? What is this pebble made of? How does a snail form its shell? To such persistent questioners, the world yields its secrets. These are the people we call "scientists" but this is a nineteenth-century word. Their forerunners were the natural philosophers. Many of them were so spellbound by the world they investigated that to follow them as far as we can, enhances our own enjoyment. Even where we cannot understand the science, their enthusiasm may rub off on us. It may even lead to faith.

The invitation given above is intended to encourage you to set out on your own exploration; to flesh out the lives outlined so briefly here, and to search out others; to look at things you have only glanced at before; to catch the excitement of discovery in both faith and science; to pray, like Annie Dillard, with your eyes open. [5]

Science is driven by a sense of wonder, an insatiable curiosity and a concern for truth. Rather like faith.

I encourage you to linger over each short chapter; perhaps not reading more than one a day. One dinner a day – nourishment. Several dinners a day – indigestion.

Notes

(1) R. Rilke, tr. J. Leishman *Requiem and Other Poems* Hogarth
(2) P. & A. Dobell *The Poetical Works of Thomas Traherne*
(3) R. Jefferies *The Open Air*
   (1), (2) & (3) are quoted in V. Gollancz *A Year of Grace* Penguin 1955 pp 88, 75 & 76 respectively.
(4) Brother Lawrence, ed. H. Martin *The Practice of the Presence of God* Student Christian Movement 1956 p 9
(5) E. Peterson *The Contemplative Pastor* Eerdmans 1993 p 67ff (Peterson uses "Praying with Eyes Open" as the title for his chapter on the inspirational writing of the American essayist, Annie Dillard).

1 – In the beginning

"In the beginning God created the heavens and the earth."[1]

In its first sentence the Bible combines aspects of science and faith. "Beginning" implies time, which science has come to measure more and more accurately. "God" is before and beyond science – a matter of faith. "Created" suggests "made to exist" rather than "came to be". It refers to the activity of God – also a matter of faith. We ourselves are an integral part of the "heavens and the earth", that amazing bundle of matter and energy which science investigates. For those with eyes to see, it is also a daily and never-ending marvel.

It is not known when humans first realised that there was a time "before the hills in order stood, or earth received her frame", but creation stories have a very distant and diverse ancestry. It seems strange that primitive peoples should ever think of their world as anything other than everlasting. Animals and plants known to their forebears shared with them the unchanging hills and valleys. The sun rose and set with unfailing regularity. Familiar patterns of stars circled the night sky. It is something of a surprise therefore to learn that the records of many early civilisations "witness to a profound and universal conviction that there was a time when both the world and the human race ... had a beginning". Further, " ... they all see the universe not as a fortuitous production, but as the

CHAPTER ONE

expression of divine purpose, in which mankind has an allotted part".[2]

For thousands of years human senses were all that was available to explore and investigate the world. Survival was the name of the game for our ancient ancestors and, through the centuries, their developing technical skills were fashioned to this end. Fire was tamed; tools and weapons invented. Grain was ground; fields irrigated; metals mined and smelted; cloth woven; clay moulded into pots; wheels, carts and sailing boats constructed. Many communities live close to these ancient technologies even today and all of us are indebted to these unknown "hewers of wood and drawers of water". However, prehistoric cave paintings and musical instruments, together with primitive religious symbols and burial practices, all suggest that, for some at least, life rose above mere survival.

When writing was invented, the world learned to store information so that it no longer depended on word of mouth for its survival. Now, records of facts could be kept and even thoughts recorded. From the practical skills of the early technologists, from the thinking of early philosophers and from the keeping of records, science has emerged and grown into the colossus that it is today.

How are we to regard science and faith? Many eminent scientists, not least Einstein and Planck, have seen no conflict between the two but regarded them rather as complementary. In a 1939 essay Einstein made the memorable statement, "Science without religion is lame; religion without science is blind."[3] It suggests that he thought the scientist who ignored religion walked one-footed in the world and that the man or woman of faith who defied observation and reason was blind to reality. Faith predates science. There were people of faith long before Thales, in the sixth century BC, began the great scientific adventure. If overnight we were to lose all the accumulated benefits of science, people the world over would be inconvenienced. The poorest would be marginally affected. For others life would be changed almost beyond recognition and many would die who now survive. However, if we were able to ask Max Planck (Ch. 37) what sustained him through the terrible tragedies of his life I suspect it would be faith rather than science.

Science and faith both have the ability to stir our sense of wonder and dramatically to affect the way in which we live. In the New Testament letter to the Hebrews we read, "By faith we understand that the universe was formed at God's command, so that what is seen was not made out of what was visible."[4] Much hinges on how we respond to, "In the beginning God created the heavens and the earth." Some will choose the science and ignore the faith. Some will choose the faith and fear the science. Or we can choose both. Among the lives featured in this book are some whose faith drives their science, and whose science feeds their wonder at God's grandeur.

Notes

(1) *Genesis* 1: 1
(2) E. Brandon *Creation Legends of the Ancient Near East* Hodder and Stoughton 1963 p 211/212
(3) A. Einstein Essay – *Science and Religion* 1939 Quoted in, ed. T. Ferris *The World Treasury of Physics, Astronomy and Mathematics* Little, Brown & Co 1991 p 832
(4) *Hebrews* 11: 3.

2 – Engrossed in the stars

THALES   c.625 - c.547 BC

"Who is like the wise man? Who knows the explanation of things? Wisdom brightens a man's face." [1]

The invention of writing made it more certain that the thoughts of one generation would pass to the next, but that could never be guaranteed. Clay tablets are durable but not indestructible. Papyrus and vellum can rot or burn. However, sometimes these early writing materials did survive long enough for the thoughts of the ancients to come down to us in copies of their writings or in other peoples' records of what they said. One of these ancients is Thales, who has been credited with being the first to think in a scientific way. In other words, he looked for facts and let them speak for themselves rather than accepting without question the authority of statements handed down by history. Science is the child of such rigorously questioning minds.

The Old Testament prophet, Jeremiah, may have been a contemporary of Thales. We can picture the Greek philosopher busy with his thoughts in the city of Miletus in Asia Minor, while the prophet, in Jerusalem, was dictating to his scribe, Baruch, "all the words the Lord had spoken to him".[2] Whether Thales' thoughts were recorded by himself or a scribe, it is unlikely in the extreme that we shall ever see their handwriting, or Baruch's, on a piece of papyrus or vellum. What a find that would be! For the Bible, faithful copying by generations of scribes has carried its sixty-six books down the centuries to the age of printing. The surviving biblical manuscripts, by their abundance and antiquity, are themselves a witness to the value placed on them over hundreds or even thousands of years.

Documents from ancient Greece have also survived, but include none written by Thales. His achievements live on in the writings of others, such as Aristotle. There is a suggestion that his practical ingenuity led to the making of a fortune in olive oil. On the other hand, Plato has a story suggesting that Thales, while pondering the stars above him, stumbled into a well. The incident sparked off the mirth of a servant, who mocked him for being so engrossed in the stars above, that he was not aware what was happening at his feet. The practical experimenter and the absent-minded dreamer are two of the stereotypes of scientists, which survive to this day.

All thinkers are influenced by the past. Egyptian geometry, together with the discoveries of Phoenician navigators and Babylonian astronomers, were available to Thales. However, he is credited with new ideas in geometry, in particular the notion of proof by induction. This is a foundation stone in science and a major reason for Thales' reputation. He made five fundamental propositions in geometry, one of which is that a circle is exactly cut in half by its diameter. The point is that he not only stated this but also showed how it could be proved.

Thales, apparently, believed that the earth was a flat disc floating on water, and that this instability accounted for earthquakes. If so, then he was perhaps the first to suggest a natural explanation of such an event rather than simply attributing it to the God Poseidon. However, for Thales, a natural explanation did not mean there was no god to whom thanks were due. It is reported that on making the discovery of his fourth proposition in geometry, he sacrificed a bull to the gods in thanks. Perhaps the seeds of the apparent conflict between science and faith are to be found in the way of thinking

CHAPTER TWO

that suggests that if something can be explained, that is all there is to it. A painting may be explained in terms of sequence of brush strokes and arrangements of pigment, but is that all? Frances Cobbe, the nineteenth-century writer on society and morals, wrote, "It is a singular fact that when we can find out how anything is done, our first conclusion seems to be, 'God did not do it'. No matter how wonderful, how beautiful, how intimately complex and delicate has been the machinery which has worked perhaps for centuries, perhaps for millions of ages, to bring about some beneficent result, if we catch but a glimpse of the wheels, its divine character disappears."[3]

We take so much for granted. I was once involved in an accident and found myself being thankful for the kindness of total strangers. It is easy enough to thank individuals on such occasions. But what about the things that simply make us glad? Whom shall we thank for the discoveries of our minds; for hands that can paint or write, manipulate apparatus or play an instrument? Whom shall we praise for the beauty of the earth, for the glory of the skies, if there is no God? The contemporary author, Philip Yancey, found that, "It is a terrible thing ... to be grateful and have no one to thank, to be awed and have no one to worship."[4] He is not alone. Much earlier, one of our poets had written, "What is gladness without gratitude, And where is gratitude without a God?"[5] We don't need, like Thales, to go in search of a bull, but we can delight in discovery and cultivate his thankful spirit. That is wisdom – and it brightens the face.

Engrossed in the stars

Notes

(1) *Ecclesiastes* 8: 1
(2) *Jeremiah* 36: 4
(3) Quoted in J. Gray *Letters of Asa Gray* Macmillan 1911 p 638
(4) P. Yancey *What's So Amazing About Grace?* Zondervan (Harper Collins) 1997 p 41
(5) ed. F.Page *The Poems of Coventry Patmore* Oxford University Press 1949 p 457.

3 – Every realm of nature is marvellous

PLATO : 427 – 348 BC
AND ARISTOTLE : 384 – 322 BC

Who came first, Socrates or Plato? When did Aristotle live, or Euclid? I suspect that the names are more familiar than the answers to the questions. We may not find it easy to say why they were famous. However, even to dabble in their writings or the literature about them, is to catch some hint of the excitement of new ideas, rather as a whiff of sizzling bacon hints at the reality in the frying pan.

Plato was born in 427 BC, probably in Athens, which was where he died aged about 80. His earlier years were shadowed by war but enlightened by Socrates (470 – 399 BC). For Socrates it had seemed "a superlative thing to know the explanation of everything, why it comes to be; why it perishes; why it is."[1] However, he became disillusioned with the 'scientists' of his day, believing them to be loud on dogma and quiet on proof. Similarly, the activities of politicians turned Plato against politics, at least as a way of life for himself. Both of them turned to more fundamental issues. Socrates concerned himself with discovering what lies at the heart of making a success of this life. What is the true good? Plato also gave himself to philosophy, but in the history of science he may be remembered more for founding the Academy (the first

Every realm of nature is marvellous

university) to which he devoted the last half of his long life. The Academy had a huge influence on mathematics in the 4th century BC and also achieved significant biological work.

Plato's own fame does not lie in original contributions to maths and science. One historian of science, G. Sarton, is quite scathing. " ... the views expressed in Plato's *Timaeus* are unscientific. He asserts this and that but he proves nothing, and his language is often as unclear as any soothsayer."[2] Perhaps Sarton had in mind statements such as the following: "Birds were produced ... from harmless, empty-headed men, who were interested in the heavens but were silly enough to think that visible evidence is all the foundation astronomy needs." Or this: "Land animals came from men who had no use for philosophy ... And the reason why some have four feet and others many was that the stupider they were the more supports God gave them."[3] Now there is an idea for the origin of species!

To Plato, philosophy was all. Observation was a diversion. For him the true realities lay in the world of pure number and perfect geometrical figures. A concern for education led him to stress the importance of mathematics in encouraging logical thought. Euclid, who flourished around 300 BC, would have delighted Plato's mathematical heart. It has been said that through Euclid every schoolchild is a student of Plato. The older you are, the truer this is likely to be and the more likely you are to have heard of Euclid's *Elements of Geometry*, the basic maths textbook for 2000 years and one of the most famous books ever written. It has even been claimed that it has exercised an influence upon the human mind greater than any other work except the Bible. The 'elements' of his geometry, which Euclid assumes are just 'there', which he takes as 'given', are points, straight lines and circles. From these simple elements he builds up his whole system of definitions and proofs for other geometrical figures.

CHAPTER THREE

Plato's most famous pupil, Aristotle, entered the Academy at the age of 17. It is too simple to say that Plato dealt in ideas and Aristotle in facts but it is true that Aristotle, unlike his teacher, accepted the senses as a route to knowledge: "There is nothing in the intellect that was not first in the senses."[4] His remarkable observations on some 500 different animals and many plants are not the sort of thing Plato would have done. Aristotle wrote, "We proceed to treat of animals without omitting, to the best of our ability, any members of the kingdom however ignoble. ... We therefore must not recoil with childish aversion from the examination of humbler animals. Every realm of nature is marvellous ... we should venture on the study of every kind of animal without distaste; for each and all will reveal to us something natural and something beautiful."[5]

This biological part of his work lay neglected and forgotten for centuries and some of the observations have been confirmed only in recent years. Indeed, in the early centuries after Aristotle's death, none of his work made an impact comparable to Plato's. Some of it was rediscovered and revived around the end of the first millennium and then through Syrian, Arabic and eventually Latin translations made its way into the thinking of Western scholars. Here it became a mainstay of the early universities in Paris, Oxford and Cambridge.

The Greeks held an early belief that the perfect form was the circle or sphere. Two hundred years before Aristotle, Pythagoras had proposed on religious grounds that the earth and heavenly bodies were perfect spheres, the latter moving in perfect circles. This was a major advance in thinking about the universe, although now we know the spheres to be irregular and the orbits elliptical. Advances in science are not always rooted in immediate observation. Pure thought and ideas of symmetry and beauty also play their part. In the autumn of 1955, the English physicist, Paul Dirac, was visiting professor at Moscow

Every realm of nature is marvellous

University. Someone asked him to state briefly his philosophy of physics. He wrote on a blackboard, "Physical Laws Should Have Mathematical Beauty".[6]

But, to return to Aristotle. Circles and spheres remained basic to his concept of the heavens. It has been suggested that medieval navigators were fearful of venturing too far lest they fell off the edge of a flat earth. However, Jeffrey Russell argues that the myth of medieval belief in a flat earth is of modern origin and that, since Aristotle's day the spherical nature of the earth has not been seriously questioned.[7] In fact, as Aristotle's reputation grew, nothing could be questioned, and his every word came to be accepted as final. It seems likely that he would have been surprised and perhaps dismayed at this rigidity, which bound scientific thinking following his death. He had built on the ideas of others. He probably expected others to build on his ideas.

In 335 BC Aristotle had founded his own school, the Lyceum. On January 16, 1997, *The Times* reported an exciting claim by archaeologists to have discovered the site in Athens. One offshoot of the Lyceum was the famous Museum in Alexandria. From there, Aristotle's legacy of discovered knowledge and wide-ranging thinking was dispersed around the world, although this did not happen smoothly or everywhere. These ancient Greeks remind us that beauty, thinking and observation all combine in the progress of science.

In Ross's *Works of Aristotle* you can read, "God, who in might is most powerful, in beauty most fair, in time immortal, in virtue supreme; for though he is invisible to all mortal nature, yet is he seen in his very works. For all that happens in the air, on the earth and in the water, may truly be said to be the work of God, who possesses the universe."[8] The quotation shows that at least some of the ancient Greeks shared the earlier Biblical idea that God is seen in his works. Not only that. The phrase, "He possesses the universe", again echoes a recurring

CHAPTER THREE

theme in the Bible: "To the Lord your God belong the heavens, even the highest heavens, the earth and everything in it."[9]

Notes

(1) N. Spurway *Humanity, Environment and God* Blackwells 1993 p 48
(2) G. Sarton *Ancient Science through the Golden Age of Greece* Dover 1993 p 450
(3) Plato *Timaeus* Penguin Classics 1965 p 121
(4) "Aristotle" Microsoft(R) Encarta(R) 98 Encyclopaedia
(5) Aristotle, ed. W. Ross *De Partibus Animalium Book 1:5* 645(a) Oxford (Clarendon) 1963
(6) R. Dalitz & R. Peierls in *Biographical Memoirs of Fellows of the Royal Society* 1986 Vol. 32 p 159
(7) J. Russell *Inventing The Flat Earth* Praeger 1991
(8) Aristotle, ed. W. Ross *De Mundo* 399b 20. *Works of Aristotle* Oxford (Clarendon) 1963
(9) *Deuteronomy* 10: 14.

4 – Enquiry into plants

THEOPHRASTUS : c.371 - c.287 BC

"Then God said, 'I give you every seed-bearing plant on the face of the whole earth and every tree that has fruit with seed in it. They will be yours for food.' "[1]

Primitive people collected a natural harvest from the wild, but long, long before the time of Theophrastus, someone had had the brilliant idea of sowing or planting deliberately. And so began the gradual domestication and improvement of crops that still continues. There must also have been a day, when, echoing Eden, somebody first grew a plant simply because he or she found it 'pleasing to the eye'. Others were collected, and later grown, for their power of healing. Many of these last feature in the Herbals which contain some of the earliest illustrations of plants and now make such fascinating and sometimes amusing reading. For example, Gerard, in his 16th century *Herball*, tells us that the juice of the leaves and roots of the common daisy "given to little dogs with milke, keepeth them from growing great."[2] Owners of large dogs who regret coming upon this gem too late, will be comforted to know that much of the information in these ancient books was handed down untested by serious experiment. The Herbals may mix science with fancy, observation with folklore, but plants have featured significantly in the story of science, as we shall see.

Early Greeks, having been first on the scientific scene, tend

CHAPTER FOUR

to become in retrospect, "fathers" of this and that. Theophrastus is known as the "Father of Botany", although Solomon, six centuries earlier, might wish to question the claim. "He described plant life, from the cedar of Lebanon to the hyssop that grows out of walls."[3] My own experience of botany suggests that their mothers may have inspired their love of plants and deserve a "Mother of Botany" title. I remember my mother telling me when a very small child that a weed in our garden was called *Persicaria*. She also told of a legend accounting for the dark spots on its leaves. "The legend is that the Persicaria was growing on Calvary at the foot of the Cross upon which Christ hung, and that the drops of blood from His wounded feet and hands dropped upon its leaves".[4] I came across this quote in *A First Book of Wild Flowers* years after my mother died. It was only then that I discovered her maiden name written inside the cover and appreciated the link with my childhood memory. Science carries a weighty debt to generations of largely unknown mothers.

Theophrastus was born at Eresus on the Greek Island of Lesbos. He became a pupil at Plato's Academy along with Aristotle, his senior by about fourteen years. This link, forged at the Academy, survived Plato's death in 348 BC, and when Aristotle died twenty-six years later, Theophrastus inherited the directorship of the Lyceum, and the library, in which were originals of Aristotle's works. It seems he was a spirited and stimulating lecturer who lived to the age of 85. When he died, his body was buried in the Lyceum garden that he had helped to create. The saying "We die just when we are beginning to live" is attributed to him. It suggests an optimistic slogan for those past middle age – "Life begins at 85". Very little is known about his long life but he wrote two botanical tomes, each, in the fashion of the day, made up of several books. Among his other writings was an early geological book, *On Stones*.

Enquiry into plants

In the *Enquiry into Plants* it is interesting to see the first inkling of what we now call ecology. He is writing about the differences plants show in appearance and habitat. " ... we must take into account the locality, and indeed it is hardly possible to do otherwise ... for there are some plants which cannot live except in wet; and again these are distinguished from one another by their fondness for different kinds of wetness; so that some grow in marshes, others in lakes, others in rivers, others even in the sea ... "[5]

Have you ever wondered why evergreens are evergreen? The vivid colours of autumn escape nobody's notice. Leaves tumble or drift to the ground to rustle underfoot or sink into soggy decay. Even my current railway timetable states that, "Between 28 September and 12 December, all trains travelling ... on Mondays to Fridays will depart 3 minutes earlier than shown on this timetable to allow for extra braking time during the leaf fall season." But what about the hollies and laurels and their kind? If you have these in the garden you will know that dead leaves *do* accumulate under them.

Theophrastus explains. "In those which are evergreen the shedding and withering of leaves takes place by degrees; for it is not the same leaves which always persist, but fresh ones are growing while the old leaves wither away. This happens chiefly about the summer solstice."[6] In other words, each evergreen leaf lasts at least a twelvemonth so there is no season when the plant is bare of leaves. The scientist does not stop short at wondering, but goes on to investigate.

It was written of the lilies of the field that not even Solomon in all his splendour could compare with them. On Tuesday, April 5, 1994, the dramatist Dennis Potter, dying of cancer, gave a final interview. "Below my window in Ross when I'm working the blossom is out in full. It's a plum tree ... it's the whitest, frothiest, blossomest blossom that there ever could be. ... The nowness of everything is absolutely wondrous ... The fact

CHAPTER FOUR

is, that you see the present tense, boy do you see it! And boy can you celebrate it."[7] The likes of Theophrastus teach us to look, *really* to look, at the lilies of the field. Why wait till we are dying to celebrate them? Why not now?

Notes

(1) *Genesis* 1: 29
(2) M. Woodward *Gerard's Herball* Spring Books 1964 p 146
(3) *1 Kings* 4: 33
(4) M. Rankin *A First Book of Wild Flowers* Melrose p 178
(5) Theophrastus, tr. A. Hort *Enquiry into Plants* Heinemann 1916 p 31
(6) Theophrastus p 67
(7) *The Guardian* April 6, 1994.

5 – Two great lights

ARISTARCHUS : c310 - c230 BC

"God made two great lights – the greater light to govern the day and the lesser light to govern the night" [1]

Faith marvels; science measures. The psalmist marvelled. "When I consider your heavens, the work of your fingers, the moon and the stars, which you have set in place, what is man that you are mindful of him, the son of man that you care for him?"[2] The scientist may also marvel and even ask the same questions as the psalmist; but because science measures, other questions will come to mind. Concerning the "two great lights", science will ask, "How great? How far away?"

It seems that the first recorded attempt to answer these questions was by Aristarchus of Samos. He realised that, at the half-moon, a line passing from his eyes to the centre of the moon would be at right angles to the light shining from the sun to the moon.

Sometimes it is possible to see both the sun and the half-moon in the sky at the same time. When this happened, Aristarchus could measure the angle between them. Using this information he worked out how many times further away the sun was from the moon than the moon was from him. He was very wrong because, although his measurement of the *angle* was less than 3° out, the distances involved are vast. This makes a huge difference to the result, as I discovered when

CHAPTER FIVE

walking on a compass bearing which was slightly wrong. The further I walked the more my route diverged from the one I intended. I did not end up where I expected. How marvellous then, for Aristarchus to think that such a measurement might even be possible at all, and then to follow it through to a method of getting *any* answer. He also realised that the sun must be very much bigger than the moon, simply because it is so much further away; a "great light" indeed.

It is said that Aristarchus thought that the earth and other known planets revolved around the sun. Some trace the seeds of such a view to even earlier thinkers but it then lay dormant for nearly two millennia. Even in his lifetime there were those who thought Aristarchus should be indicted on a charge of impiety, "for putting the earth in motion". It is intriguing that from the earliest flickerings of science any apparent contradiction of current dogma has been considered impious.

This evening, as I write, there is one crisp, crescent moon in a clear sky. How different things might have been. It might have been bigger or much smaller; nearer to earth or more remote. Imagine a moon whizzing round the earth several times a night. We might have had many moons, or none. In 1969 Neil Armstrong left his footprints in the dust of our moon's barren surface, and yet how little most of us know about it. If you doubt that last statement, take your next opportunity to look at a full moon and see how many surface features you can name. Most of us have little idea where the moon will appear in the sky or what it will look like in two weeks' time, let alone how we might pinpoint the dates and times of eclipses over the next century.

The sole surviving work of Aristarchus, *On the Sizes and Distances of the Sun and the Moon,* began with six assumptions, the first of which is, "That the moon receives its light from the

Two great lights

sun."⁽³⁾ This was nearly nineteen hundred years before Galileo but in the latter's lifetime there were people who still believed that moonlight was moonlight, and not reflected sunlight. One such man defied the astronomers of the time by publishing a statement that, "the moon does not receive its light from the sun but is brilliant by its own nature." This same man quoted the Bible to try to give authority to his wrong statement. This upset Galileo who wrote, "He supports this fancy (or rather thinks he does) by sundry texts from Scripture which he believes cannot be explained unless his theory is true;" and he adds with more than a touch of irony, " ... yet that the moon is inherently dark is surely as plain as daylight."⁽⁴⁾

Galileo was very concerned that respect for the Bible should not be lowered by quoting it against proven facts, as above, or even against theories, which might later come to be proved. Two hundred years later, Charles Babbage, Lucasian Professor of Mathematics at Cambridge from 1828 to 1839, was giving similar warnings. This time it concerned the age of the earth but his fear was the same, namely that "the authority of scripture will be gradually undermined by the weak though well-intentioned efforts of its friends in its support."⁽⁵⁾

When Aristarchus was investigating the heavens much of today's Bible was still unwritten. The books making up the Old Testament as we have it today were finally settled at a council of Jewish Rabbis and scholars near the end of the first century AD. It was not until the Easter letter of Athanasius in AD 367 that we have the complete list of books that we find in a modern New Testament. Even so, by the time Galileo turned his telescope to the stars, the Bible as we know it had already circulated for more than 1200 years. His criticisms were not directed at the Bible but at some of its interpreters. He quoted with approval Cardinal Baronius who, referring to Scripture, suggested "That the intention of the Holy Ghost is to teach us how one goes to heaven, not how heaven goes."⁽⁶⁾ Galileo

CHAPTER FIVE

himself was more specific: "I believe that the intention of Holy Writ was to persuade men of the truths necessary for salvation, such as neither science nor any other means could render credible, but only the voice of the Holy Spirit."[7]

Notes

(1) *Genesis* 1: 16
(2) *Psalm* 8: 3,4
(3) ed. C. Gillispie *Dictionary of Scientific Biography* Charles Scribner's Sons NY 1981 Vol. 1 p 248
(4) S. Drake *Discoveries and Opinions of Galileo* Doubleday Anchor 1957 p 190
(5) ed. M. Campbell Kelly *The Works of Charles Babbage* Vol. 9 *The Ninth Bridgwater Treatise* 2nd ed. Pickering 1989 p 17
(6) S. Drake p 186
(7) D. Sobel *Galileo's Daughter* Fourth Estate 1999 p 65.

6 – The proof will carry conviction

ARCHIMEDES : c.287 – 212 BC

You probably, like me, have never had the chance to handle a gold ingot. Apparently those who have done so are surprised to find it seemingly glued to the table when they try to pick it up. The surprise is that it has so much weight for its size. Bear in mind through what follows, that lumps of a pure metal having the same weight will also have the same size, or volume.

If you know anything at all about Archimedes it will be that he dashed naked through the streets of his native Syracuse in Sicily yelling: "Eureka!" or, in English, something like, "I've found it!" You may know the story (and some claim that it is only a story) but have no idea what he had found or why he was so excited. Apparently King Hieron wanted to know if his crown, said to be of pure gold, contained any silver, and had asked Archimedes how he could find out.

We can imagine that as Archimedes sank, with a sigh of satisfaction, into the bath that day the answer came to him. Hence his joy as he sped home. Everyone knew that the water level rose when one got into a bath. A lot of people could work out the volume of regular objects like cubes and cylinders. It was Archimedes' genius to realise that if he measured the length and breadth of his rectangular bath and the amount by which the level rose when he submerged, he could calculate the

CHAPTER SIX

volume of water displaced, which obviously had the same volume as his own body. What was true for his body, he realised, must be true for the crown, or any other irregularly shaped object. So, if he now immersed the crown in a vessel of water, he could measure its volume. All he had to do then was to ask the king for a lump of gold with exactly the same weight as the crown. Each was then immersed in water and the rise in level for each noted. If this rise, and therefore the volume, was the same for each, then the crown was pure gold. Whether or not it was, we don't know!

In one of Archimedes' letters, I was struck by a similarity with the way Luke begins *Acts* (*The Acts of the Apostles*). In the letter we read, "Archimedes to Dositheus; Greeting. On a former occasion I sent you the investigations which I had up to that time completed ... "[1] *Acts* opens with: "In my former book, Theophilus, I wrote about all that Jesus began both to do and to teach ... "[2] How marvellous that the words of these ancient writers (albeit in translation) have survived the centuries. Did Luke read Archimedes? Did he understand him? I have to confess that generally, when I tried to read Archimedes' Propositions, my eyes glazed, and circles, segments, arcs, radii, spheres and centres of gravity tumbled together into an impenetrable gobbledegook. Better minds than mine, however, delight in the step by step "thus", "hence" and "therefore" as he works his way meticulously to some new insight in arithmetic, hydrostatics, mechanics, plane or solid geometry, or astronomy. Such was the breadth of Archimedes' interests. Some consider him perhaps the greatest mathematical genius the world has ever seen.

Floating things *on* water was just as fascinating to Archimedes as immersing things *in* it. One does not have to know very much about floating to realise that to make ships out of iron seems to fly in the face of common sense. Archimedes prepared the way for such vessels centuries earlier in his work

The proof will carry conviction

On Floating Bodies. It consists of nine propositions in the first book, and ten in the second.

Proposition 5 is the "Archimedes' Principle" which some of us vaguely recall from schooldays. The gist of it is that a floating body displaces its own weight of fluid. So, if you float a small bowl in a larger one, which is already full to the brim, the water that overflows weighs the same as the small bowl. If you now put something into the small bowl, but so that it still floats, it will sink deeper and more water will overflow, weighing just the same as whatever you put in. For this particular proposition I relished the fact that I *could* understand his proof that followed, and enjoy a faint echo of the delight he must have felt in making it.

During his lifetime Archimedes was famous for various mechanical inventions. Perhaps you have seen water being raised from a stream into an irrigation ditch by turning a screw within a cylinder. The same kind of device is also used to move grain, from silo to lorry, for example. It is the Archimedes' screw. Its inventor died by the hand of a Roman soldier at the fall of Syracuse, thrust through as he pondered mathematical figures in the sand.

The Bible often uses "the sand on the seashore" to represent a vast, uncountable number. "I will surely make you prosper and will make your descendants like the sand of the sea, which cannot be counted."[3] Did Archimedes read Genesis? Whether he did or not, this kind of statement posed a mathematical challenge which he was unable to resist. He calculated, for King Hieron's successor, not only the grains of sand on the seashore, but the number in a sphere the size of the then-known universe. He concludes; "I conceive that these things, King Gelon, will appear incredible to the great majority of people who have not studied mathematics, but to those who are conversant therewith and have given thought to the

CHAPTER SIX

question of the distances and sizes of the earth, the sun and the moon and the whole universe, the proof will carry conviction."[4] This concern with proof that will carry conviction is the backbone of science throughout its history.

Notes

(1) T. Heath *The Works of Archimedes* Cambridge University Press 1897 p 1
(2) *The Acts of the Apostles* 1: 1
(3) *Genesis* 32: 12
(4) T. Heath, p 232.

7 – Tireless diligence in calculating

PTOLEMY : c.100 – c.175

January 30, 1996: "Life on Mars more likely than not". August 8: "Clinton hails discovery of life on Mars". December 23: "Doubts over life on Mars". Sensations make news, even if they are unproved or do not happen. More recently (February 2004) contact has been lost with Beagle 2, the lander sent by the UK-led mission to Mars. Happily, NASA's Mars rovers, 'Spirit' and 'Opportunity' landed safely and are now at work among the red dust and rocks of the Martian surface. Science depends on events like these to hit the headlines, whereas much of the real science is confined to the circle of those who read the journals in which it is published. However, thanks to the science correspondents, some of it does come to light.

To find out how much, I kept a scrapbook of science headlines taken from a daily newspaper over the course of a year. It is from this scrapbook that the examples above are drawn. It makes fascinating reading, and reveals discoveries (and doubts) in astronomy, physics, medicine, genetics, conservation, and so on, sometimes on an almost daily basis. Contrast this with the fact that the Ptolemaic system explaining the movement of the heavenly bodies survived more or less intact for 1500 years. In our "here today, gone tomorrow" world such stability of ideas seems beyond belief.

Hardly anything is known of Ptolemy the man. Different

CHAPTER SEVEN

versions of his name suggest Greek ancestry and Roman citizenship. It is more than likely that he worked in Alexandria, and observations of the heavens recorded in his most famous book, *Almagest*, show that he was busy between 127 – 141. It is suggested that he was probably born around the year 100, and lived to be about 70.

Like many early works, *Almagest* is in the form of several "books", thirteen in this case. Book 3, for example, concerns the sun, and Book 4, the moon. Books 7 and 8 are about the fixed stars. *Almagest* has been described as "a manual covering the whole of mathematic astronomy as the ancients conceived it."[1] Of course, Ptolemy was not starting from scratch and his book is more encyclopaedia than original work. Eudoxus, as early as the 4th century BC, made a serious attempt to use a mathematical model to describe the movement of the heavenly bodies. Two hundred years before Ptolemy, Hipparchus used mathematics to explain observations of the early astronomers.

None the less, it was Ptolemy who drew it all together into a system, which suffered no serious challenge until the sixteenth century. In his system, the earth was the stationary core of the universe. Around it, carried by a succession of revolving spheres were the moon, the sun and the five "wandering stars", the then-known planets – Mercury, Venus, Mars, Jupiter and Saturn. Finally, at an immense distance, was the outermost sphere carrying the "fixed stars", so called because, unlike the planets, they showed no movement relative to each other. Because the planets followed such peculiar courses as seen from the earth a complex series of circles, called epicycles, was introduced to try to explain these movements.

Some have described Ptolemy as a hack or even a fraud but to most his fame seems well founded, not only in *Almagest*, but also in works on geography (an attempt to map the then-known world), optics (theory of vision) and music. George Joachim

Tireless diligence in calculating

Rheticus, a sixteenth-century Professor of Mathematics at Wittenberg, was fulsome in his praise. "For Ptolemy's tireless diligence in calculating, his almost superhuman accuracy in observing, his truly divine procedure in examining and investigating all the motions and appearances, and finally his consistent method of statement and proof cannot be sufficiently admired and praised by anyone to whom Urania is gracious."[2]

Ancient books like the Bible or *Almagest* seldom came directly into the hands of readers in Western Europe. When Ptolemy originally wrote in Greek, his title was *Mathmatike Syntaxis*, 'a mathematical (or astronomical) compilation'. It came to be known as 'the great, or greatest, compilation'. The Caliph al-Mamun had the book translated into Arabic in 827 under the title *Al Majisti*. The eighth-century Moorish invasion of Spain had prepared the way for such translations to be carried into southern Europe. When the twelfth-century Archbishop Raymond set up a college of translators, at Toledo, in Spain, these ancient texts became available to a wider circle of readers.

Among the translators the best known is Gerard of Cremona (1114–1187), famous for his translations of works by Aristotle, Euclid, and many Arab authors. Gerard translated Ptolemy's work into Latin in 1175 as *Almagesti* or *Almagestum*. By this route *Almagest* began to influence western thinking a thousand years after its author's death. It was the book on which the European astronomers, Copernicus, Tyco Brahe and Kepler cut their teeth, but there was a new spirit in the air and they did not feel bound by its authority. Their thinking eventually rendered it out of date, and the Ptolemaic system passed into history. Even so, *Almagest* remains a source of ancient astronomical observations. Copies still exist. At least one has appeared on television, being handled with appropriate awe.[3]

CHAPTER SEVEN

There is food for thought in realising how much we owe to the care and patience of early scribes and translators, for the survival of all ancient books.

Notes

(1) ed. C. Gillispie *Dictionary of Scientific Biography* Charles Scribmer's Sons NY Vol. 11   p 187
(2) tr. E. Rosen, ed. A. Evans *Three Copernican Treatises* Columbia University Press 1939 p 131 (Urania, Muse of Astronomy)
(3) Channel 4 *Byzantium: The Lost Empire* 20/9/97.

8 – One day in seven

Subrahmanyan Chandrasekhar, 'Chandra' to his friends, was a twentieth-century astrophysicist. He won the Nobel Prize for Physics, jointly, in 1983, for his theory about how massive stars behave in their later stages. He is quoted as follows: "One of the unfortunate facts about the pursuit of science the way I have done it is that it has distorted my personality. I had to sacrifice other interests in life – literature, music, traveling. I've devoted all my time, every living hour practically, to my work. I wanted to read all the plays of Shakespeare very carefully, line by line, word by word. I have never found the time to do it. I know I could have been a different person had I done this. I don't know if regret is the right word for what I feel. But sooner or later one has to reconcile these losses. One has to come to terms with oneself. One needs some time to get things in order."[1] His obituary in the Royal Society Memoirs indicates a fuller life than this quotation might suggest, but we do all need "time to get things in order".

The creation narrative in *Genesis* describes six days of creative activity followed by a day of rest. "By the seventh day God had finished the work he had been doing; so on the seventh day he rested from all his work." (The Hebrew verb translated 'rested' here, gives us the word 'Sabbath'.) "And God blessed the seventh day and made it holy ... "[2] Holiness and wholeness have similar roots in English. There is the sense of something being deliberately set apart. We read in *Exodus* 20: 8, as one of the Ten Commandments, "Remember the Sabbath day by keeping it holy". The Commandments are

CHAPTER EIGHT

often thought of in a negative way. What a mistake! They are, in fact, like the boundaries set by loving parents for their child's safety and well-being. How easily we reject a life of freedom within such boundaries in favour of all kinds of slavery offered by independence. Of course, they never seem like slavery to begin with. Every self-indulgence and most addiction begins as a freely chosen adventure.

The Sabbath also sometimes attracts negative comment. But Jesus said, "The Sabbath was made for man, not man for the Sabbath."[3] He also named himself "Lord of the Sabbath" and said, "I have come that they may have life, and have it to the full."[4] Could anything be more positive? Ezekiel reminds us that Commandments and Sabbath are both God's gifts to us. "I gave them my decrees and made known to them my laws, for the man who obeys them will live by them. Also I gave them my Sabbaths as a sign between us, so they would know that I the Lord made them holy."[5] In the Sabbath we have a wonderfully rich resource for our well-being.

It seems that, for a while, the early Church celebrated both Saturday, the Jewish Sabbath; and Sunday, the "day of resurrection". However, the resurrection of Jesus was of such overwhelming significance that Sunday, "The Lord's day", became more important for Christians. For about two millennia we have kept this tradition of meeting together to celebrate on Sunday. In this way, Sunday came to be called our Sabbath. It is a day for worship. It includes the songs and hymns that we sing, but in his letter to the Romans, Paul gives worship a much wider meaning. He urges us to offer our bodies in service and to renew our minds. It is also a day for rest and reflection; a pit stop for the next lap of the race. It was written of Michael Faraday, "I think that a good deal of Faraday's week-day strength and persistency might be referred to his Sunday Exercises. He drinks from a fount on Sunday which refreshes his soul for a week."[6]

One day in seven

Asa Gray, an American botanist, was another scientist who took seriously the challenge to "remember the Sabbath". It was his rule to rest on Sunday. It seems too, that he made a habit of church attendance wherever he found himself. In London on January 27, 1839, he wrote: "I attended St John's Chapel today and have reason to be glad that I did so, for I heard a most excellent sermon in the morning from *Psalm* 103: 10 - 12. Mr. Noel (Baptist Noel) is a most simple, winning preacher, and his sermon was the most thoroughly evangelical and earnest I ever heard from an Episcopal pulpit."[7]

If the setting apart of one day in seven is so significant, it is surely worth giving the matter serious thought. Each of us needs to explore how we can best "remember the Sabbath". Modern patterns of work may pose a problem. Some people are on shift work; some may have no choice but to work on Sundays. Archimedes saw a problem as a challenge; and perhaps a scientific attitude can help here. Faced with a problem, the scientist will investigate, explore, experiment, until a solution is found. We too may need to persevere in exploring ways of carving our own Sabbath out of a week that appears to be a solid slab of other activities. Perhaps we can help others to do so.

In the *Genesis* account of creation the activity of God is not the whole story. Rest is there too. As our heavenly Father rested, so He encourages each of us to, "Be still, and know that I am God".[8] It was a priority for Jesus. "Very early in the morning, while it was still dark, Jesus got up, left the house and went off to a solitary place where he prayed."[9] Rest is not sloth. Jesus was criticised for actively doing good on the Sabbath, and when he walked through the cornfields with his disciples, he probably joined them in rubbing the ears of wheat in his hands and chewing the nutty grain. That too drew from his enemies the accusation of Sabbath-breaking. Because Jesus had nourished his relationship with his Father in the

CHAPTER EIGHT

solitary place he was able to shun a Sabbath of rules and regulations. He was free to worship God in the temple, to go about looking for opportunities of doing good, or to celebrate with his disciples the sheer goodness of what God had done in creation. Shall we join him?

Notes

(1) ed. T. Ferris *The World Treasury of Physics, Astronomy and Mathematics* Little, Brown & Co 1991 p 613
(2) *Genesis* 2: 2
(3) *Mark* 2: 27
(4) *John* 10:10
(5) *Ezekiel* 20:11
(6) ed. C. Gillispie *Dictionary of Scientific Biography* Charles Scribner's Sons NY 1981 Vol. 4 p 527
(7) J. Gray *Letters of Asa Gray* MacMillan 1911 p 122.
(A portrait of Baptist Noel (1799 - 1873) hangs in Regent's Park College, Oxford, he having become a Baptist in 1848. He was twice President of the Baptist Union.)
(8) *Psalm* 46:10
(9) *Mark* 1: 35.

9 – A sun-centred universe

NICOLAUS COPERNICUS : 1473 – 1543

If you were asked for evidence that the Earth is a sphere, you might well show a photograph taken from space. Supposing the request had been made before the space age; before the invention of photography and telescopes. What then? And on what grounds do you believe that the Earth goes round the sun? Surely, it is *obvious* that the reverse is true. We see the sun rise every fair morning of our lives, and each clear evening it sinks below the western horizon. For thousands of years this common-sense view prevailed and it helps us to realise what a huge leap of imagination it took to consider even for a moment that it might be the Earth that moved. So huge was it that we are now jumping nearly 1400 years from Ptolemy and his summing up of the achievements of the Greeks, to Copernicus and one of the most dramatic events in human thinking.

It should not be assumed that nothing happened during these years. Science and mathematics flourished in the Arab community, both in translation from the Greek and in original work. India and China were at the same time advancing maths and technology. Even in Europe, from the twelfth century onwards, there were some notable individual thinkers, such as Hildegard of Bingen (1098 – 1179) in Germany, Robert Grosseteste (c1168 – 1253) and Roger Bacon (c1219 – c1292) in England and Jean Buridan (c1295 – c1358) in France. When Buridan died, a century was still to pass before the printing

CHAPTER NINE

press transformed the spread of knowledge by printed books. From the birth of Copernicus onward, scientific ideas could take wing, at first in Europe but eventually around the globe.

The idea that the Earth orbited the sun has been attributed to Aristarchus, among others, 1700 years before the birth of Copernicus. Most people at that time and since had felt the idea to be ridiculous. It was because of such disagreements that Copernicus resolved to seek a new theory of the heavenly bodies.

Nicolaus was born at Torun, Poland, on February 19, 1473, the youngest of four children. The father died in 1483 and Nicolaus's uncle, by then the Bishop of Warmia, took charge of his education. He entered university at Krakow and then crossed the Alps to Italy for further studies, including medicine, at Bologna and Padua. A medical prescription noted during his studies reads, "Take two quarts of sublimed wine, four drachmas of figs, five drachmas respectively of cinnamon, cloves and saffron. Mix and strain into a clean vessel. Use freely and without restraint. If God wills, it will help."[1] We are left in the dark as to what it would help or how, and there seems to be more faith in God than in the prescription. Nicolaus had arrived in Bologna in 1496 and it was perhaps here, by the lectures of Domenico Maria Novara, that his interest in astronomy was first kindled. The enthusiasm of one generation sparks the genius of the next. Astronomy may have seemed to Copernicus to have more certainty about it than prescriptions offered in hope. However, he did not abandon medicine.

In 1497 the uncle succeeded in getting his nephew elected a canon of the cathedral chapter of Frombork, on the Baltic coast. The appointment did not interrupt his studies. On return to Poland in 1503, he practised as a much-loved physician, and worked with his uncle as secretary and adviser. The Bishop's palace at Lidzbark became his home. When his

A sun-centred universe

uncle died nine years later, Copernicus moved to Frombork and eventually settled there permanently.

Whatever else he was busy at, he also found time to develop the astronomical ideas on which his fame rests. His first proposals for a sun-centred universe were probably penned while at Lidzbark. They were certainly circulating in manuscript form among his friends by May 1514. Here he stated that, "The centre of the Earth is not the centre of the universe" and, "We revolve around the sun like any other planet."[2] These were radical ideas and he was only persuaded to publish in full right at the end of his life. A copy of the first edition was brought to him on May 24, 1543, the day he died. Such exquisite sense of timing put him beyond all personal consequences of publication. The book was *De Revolutionibus Orbium Coelestium, On the Revolutions of the Heavenly Bodies.* It dramatically recast the model of the universe.

That the book was published at all owes much, it seems, to the efforts of a young Protestant Professor of Maths who arrived on the Catholic astronomer's doorstep on a spring day in 1539 and stayed for two years. Reports of the novel ideas from Frombork had arrived in Wittenberg and reached the ears of George Joachim Rheticus, the professor, and Martin Luther, the reformer. They reacted very differently. The former "was filled with great eagerness to become acquainted with the new system",[3] and set off forthwith to Frombork. Luther's comment was "This fool [or perhaps simply 'fellow'] seeks to overthrow the whole art of astronomy! But, as the Bible shows, Joshua commanded the Sun to stand still, but not the earth."[4] We may believe that the world remains profoundly in debt to Luther for pointing the Church back to the Bible, and yet at the same time acknowledge that he made mistakes. This may not have been one of his more serious ones, for he was probably representing current opinion on both astronomy and the Bible.

In *The Revolutions* Copernicus put forward what he believed

CHAPTER NINE

to be the truth. He retained many features of the Ptolemaic system. The heavenly bodies still moved in perfect circles in a spherical, finite universe bounded by the sphere of fixed stars. However, he moved the sun to the centre and earth took her place among the orbiting planets, while retaining the moon as her satellite. In the drawing Copernicus made to illustrate all this the line showing the orbit for Jupiter betrays what looks like a slip of the compasses. Anyone who has tried using a pair of compasses to draw a circle will know how easily this happens. I was delighted to find a portrait by Bacciarelli showing Copernicus using such compasses.[5] Commenting on his proposals, Copernicus wrote, "All these statements are difficult and almost inconceivable, being of course opposed to the beliefs of many people. Yet as we proceed, with God's help, I shall make them clearer than sunlight, at any rate to those not unacquainted with the science of astronomy."[6]

These "difficult statements" gradually won approval and the proposal that it is the earth that circles the sun, rather than vice versa, did eventually come to be accepted. This was in spite of common sense opinion. It was also in spite of what many in the Church hierarchy understood the Bible to say on the subject. Luther's comment on Copernicus, if he has been reported correctly, carries a warning to all of us. Dogmatic statements about astronomy, or any other area of science, based on a particular understanding of the Bible, have a tendency to backfire. Certainly we may argue for our interpretation of Scripture provided it is done with integrity and not from a partisan spirit. Sadly, the Church, or groups of Christians within it, have too often insisted on the acceptance of a particular interpretation on pain of exclusion or worse. However high the motive, such insistence has never in the long run served the Church or the Bible or the cause of Christ. The Bible has such a rich history of speaking powerfully into people's lives that we need to let it speak for itself. Our

A sun-centred universe

attempts to insist on a particular interpretation as 'the sole truth' or 'self-evident truth' are like offering crutches to an Olympic runner. They are not needed. In the words of Galileo, " ... it is not in the power of any created being to make things true or false, for this belongs to their own nature and to the fact."[7]

It is not the statement that something is true that makes it so. Vindication of Copernicus's statement had to wait for the build up of evidence. The Bible is not a scientific statement and is therefore not answerable as a whole to that kind of evidence. However, it is rooted in history and makes statements about which science and especially archaeology may have something to say. They need to be taken seriously in our efforts to understand what the Bible means. It is not so much a question of " ... letting science determine how the Bible ought to be interpreted" as understanding where science is relevant and where it is not.

The Bible displays a many-faceted truth that has stood the test of time for hundreds of years. We need to come to it with a mind genuinely open to the whole thrust of Scripture rather than looking for texts which confirm our own thinking. Christians differ widely in the importance they give to verbal agreement on doctrine and the meaning of passages in the Bible. Imagine that every Christian accepted that the bedrock of any Christian community was following Jesus rather than agreement among the members of that community about what the Bible says. The fall of barriers would make the end of the Berlin Wall seem like the crumbling of a sandcastle.

CHAPTER NINE

Notes

(1) H. Bietkowski & W. Zonn, tr. D. Heaton-Potworowska   *The World of Copernicus*   Arkady   1973   p 88
(2) tr. E. Rosen, ed. A. Evans   *Three Copernican Treatises*   Columbia University Press   1939   p 58/59
(3) tr. E. Rosen, ed. A. Evans   p 4
(4) H. Bietkowski & W. Zonn   p 161
(5) H. Bietkowski & W. Zonn   Frontispiece
(6) N. Copernicus, tr. E. Rosen, ed. J. Dobrzycki   *Complete Works*   Macmillan – Polish Scientific Publishers   1972   Vol. 2   p 21
(7) S. Drake   *Discoveries and Opinions of Galileo*   p 210.

10 – Plants which caught his eye

LEONHARD RAUWOLF : 1535 – 1596

I remember with enormous pleasure my introduction to field botany in the magnificent countryside around Bangor in North Wales in the early 1950s. I was already interested in plants but knew only small areas of southern England, so I became fascinated by species completely new to me in that unfamiliar Welsh landscape. It was there that I had my first sight of sundew, and butterwort, both small but entertaining plants, which catch and feed on insects. There were the bold yellow heads of globeflower and, in the mountain bogs, the white tufts of cottongrass. Woodlands were sometimes carpeted white with ramsons. There were saxifrages I'd never seen before.

I began to learn to recognise the plants associated with oakwood and water meadows, with limestone rock and sand dunes; to realise there were patterns in vegetation, which could be related to differences in soil and climate. Even now, this knowledge enriches the simplest walk. Another whole new world was revealed by Professor Paul Richards, prone among the lush moss flora in the Aber valley. With the help of nothing more complicated than a high-powered magnifier, he showed that there is far more to mosses than meets the eye. I have often thought that a good definition of education would be "contact with enthusiasm" and certainly the enthusiasm of staff in the botany and agricultural botany departments of the

CHAPTER TEN

university rubbed off on me. I am sure they would have relished Rauwolf's company.

Leonhard was born into a Lutheran family on June 21, 1535, in Augsburg, Germany. In 1517, Luther had proclaimed his 95 theses in the Castle Church at Wittenberg, so Leonhard's parents must have been among the first generation to adopt the reforming monk's ideas.

At the age of 25 Leonhard enrolled at Montpellier University to study medicine, which included botanical studies. While walking in the countryside of Languedoc and Provence he began collecting plants which caught his eye. He pressed and dried them carefully to begin the large collection, or herbarium, which he then added to throughout his life. In 1690 the herbarium was bought by the University of Leiden, which confirms that it remains one of their most valuable holdings, treasured to this day.

His life's work was as a physician, mainly in his home town of Augsburg, but that did not stop him maintaining a botanic garden and enjoying the company of other botanists. They named him "Dasylycus" or "Shaggy Wolf" for reasons we can only guess at.

In 1565 he married Regina Jung. We do not know if the marriage was a happy one but eight years later, he felt able to leave his wife behind when he jumped at the opportunity to take the field trip of a lifetime. His brother was involved with a merchant firm, and it was probably the firm's interest in the discovery of new drugs that suggested a botanical expedition to the Middle East. It had been Rauwolf's dream "to discover and to learn to know the beautiful plants and herbs described by Theophrastus, Dioscorides, Avicenna and others in the location and places where they grow."[1] On May 18, 1573, the dream came true as he set off on nearly three years of travel. He did not return to Augsburg, and Regina, until February 12, 1576.

Plants which caught his eye

His travels through Aleppo to Baghdad and back were full of adventure. However, in spite of losing a precious notebook to a sabre-wielding Arab, he returned with fascinating and detailed plant descriptions. Boarding the "St Matthew" for the return voyage to Venice was not the end of his perils, for the journey proved hazardous in the extreme. His account of the ferocious storm they encountered is reminiscent of the one that left Paul shipwrecked on Malta.[2] Rauwolf records that, during a lull, "The seamen fell upon their knees and began to pray to the patron saint of their choice ... most asked St Nicholas to intercede for their safety ... All of this amazed the Lutheran Rauwolf who could not honestly tell whether he 'was more astonished at their prayers or at the tempestuousness of the sea.' No one had sought the assistance of Christ ... to whom 'Rauwolf had turned when the waves had covered the ship, just as the disciples had once wakened him during a storm'."[3] It was two centuries after Rauwolf's birth that a group of Moravian Christians, at prayer during an Atlantic storm, so impressed John Wesley that he longed for the confidence in Christ that they showed.

Leonhard's travel book, based on the diary of this whole trip, reveals not only his dramatic adventures but also his devout Lutheran faith. He had made a point of visiting the land of Canaan to see some of the places mentioned in the Bible. He thought it would help him to "make my own, by faith and firm confidence, Christ our Lord himself together with his heavenly gifts and treasures, as he has manifested himself in the Holy Scriptures."[4] Rauwolf was dismayed at the way some of his fellow pilgrims relied on indulgences to secure their souls. "He had not come on this pilgrimage to obtain indulgences, ... 'because all these things are directly contrary to Scriptures ... He expected remission of sin no other ways but only in the name and for the merit of our Lord Jesus Christ'."[5] While his companions were burdening themselves with bits of

CHAPTER TEN

stone that they knocked off various buildings to take home for their friends, Rauwolf was remembering how Christ had come to "make us free of the heavy burden of our sins."[6]

Rauwolf was an early and successful field botanist. Very much later a Genus of tropical shrubs was named after him and, well-loved doctor that he was, he would have been pleased to know that one of them, *Rauwolfia serpentina*, yielded the drug reserpine. It may be used to lower high blood pressure and to relieve anxiety. Medicine, botany, travel; all were important to him, but it is easy to see where his first priority lay. He probably had his parents (and, indirectly, Luther) to thank for faith rooted in the Bible; a faith well established before the storm broke. Why do so many of us wait for the storms of life before putting our trust "in the name and the merits of our Lord Jesus Christ"? Rauwolf's phrase may read quaintly today, but who Jesus is, and what he has done for us can be found in a modern New Testament at least as readily as in a sixteenth-century one.

Notes

(1) ed. C. Gillispie *Dictionary of Scientific Biography* Charles Scribner's Sons NY 1981 Vol. 11 p 311
(2) *Acts 27*
(3) K. Dannenfeldt *Leonhard Rauwolf* Harvard University Press 1968 p 172
(4), (5), (6) K. Dannenfeldt p 161.

11 – The secrets of magnetism

WILLIAM GILBERT : 1544 – 1603

It is a dry winter's night. I get into bed and notice a spark as I turn back the bed cover. I brush my hand across the cover deliberately and am rewarded by a more lively display. I try again, to see just how bright a spark I can make – and then settle to sleep. If William Gilbert had ever seen sparks like these you may be sure his interest would not have faded like mine.

He was the first to record systematic experiments in what we now call static electricity. He recorded the results of rubbing, with a dry cloth, amber and many other things; precious or semi-precious stones, sealing wax, glass, sulphur and so on. He found they all attracted small pieces of almost anything and gave them the name 'electrics', after the Greek word for amber. By his use of this name and by his experiments, Gilbert may be said to have established electricity as a separate field of study. He investigated how drops of water were affected by his electrics. You can get the idea by rubbing a plastic ruler with a dry cloth and then holding it near a dripping tap. He noticed that the weather made a difference and wrote, "Weather affects an electric's action, which is best when the Atmosphere is thin, and there is a north wind or, in England, an East one; during South winds and in damp weather, electrics behave poorly or do not attract at all."[1] In

CHAPTER ELEVEN

his famous book on magnets, *De Magnete*, all this electrical work was merely by way of introduction, to show what a difference there was between "the amber effect" (i.e. static electricity) and magnetism.

William Gilbert (or Gilberd) was born in Colchester and by 1570 was a senior fellow of St John's College, Cambridge. He never married. It was written of him that, "He had the clearness of Venice glass without the bitterness thereof, soon ripe and long lasting in his perfections. He commenced doctor in physick and was physician to Queen Elizabeth, who stamped on him many marks of her favour, besides an annual pension to encourage his studies. ... His stature was tall, complexion cheerfull, an happiness not ordinary in so hard a student and retired a person."[2]

Gilbert's experiments were fitted into time squeezed from his busy life as a doctor. Although he investigated both electrical and magnetic effects, neither of these was newly discovered. Both had been known to the ancients in Greece and China. Shen Kua of China, in his *Essays from the Torrent of Dreams,* 1086, provides the earliest known record of a magnetic compass being used for navigation at sea, but by then it had already been known for a thousand years at least, that a piece of lodestone (magnetic iron ore) could indicate the north pole.[3] Even the experimental approach was not entirely new, for Gilbert himself refers to a letter on magnetism by Peter Perigrinus from the thirteenth century, "one of the most impressive scientific treatises of the Middle Ages".[4]

William Gilbert died on November 30, 1603, not many months after the Queen. On December 6 of that year Sir Michael Hicks wrote to the Earl of Shrewsbury; "I heard as I was writing ... Doctor Gilbert, the physician, is dead, who was my neighbour at St. Peter's Hill. He was a learned physician and an honest. The sickness is greatly decreased in London, and the citizens do return daily in great numbers."[5] Some presume that he died of the plague.

For science to progress, careful recording is as important as the experiments themselves so that they can be repeated by others and the results confirmed or refuted. In this, Gilbert was a true scientist, for, in *De Magnete*, he wrote, "He who wishes to try these (our discoveries and experiments), let him handle the bodies, not carelessly and ineptly – but with skill, and properly; nor let him (when things do not succeed) ignorantly accuse our discoveries. For nothing is brought forth in this book which has not been investigated and made completely certain, performed, done in our presence."[6] His confidence in the fact that he was beginning to unravel the secrets of magnetism in no way, for him, diminished the Author of creation. He wrote, "By the wonderful wisdom of the Creator therefore forces were implanted in the earth."[7] He was referring to magnetic forces.

Gilbert identified the earth itself as a giant lodestone and he used a small spherical magnet thoroughly to explore all the movements due to magnetism. His book was known to Galileo who, despite wishing Gilbert had been a better mathematician, wrote concerning him, "I extremely praise, admire and envy this author. ... I think him moreover worthy of extraordinary applause."[8] We can applaud Gilbert because he allowed simple observations to lead him into a lifetime's fascinating study and open up whole new fields of investigation. There lies the difference between the scientist and others, for whom the same observations might stir a passing interest or none at all.

CHAPTER ELEVEN

Notes

(1) D. Roller   *The "De Magnete" of William Gilbert*   Menno Hertzberger, Amsterdam   1959   p 114
(2) S. Thompson   *Gilbert, Physician*   Note on the 300th anniversary of his death   Whittingham, Chiswick   p 31
(3) D. Roller   p 33
(4) ed. C. Gillispie   *Dictionary of Scientific Biography*   Charles Scribner's Sons   1981   Vol. 10   p 537
(5) D. Roller   p 90
(6) D. Roller   p 112
(7) W. Gilbert, tr. P. Fleury Motteley   *The Loadstone and Magnetic Bodies*   Quaritch   1893   p 328
(8) D. Roller   p 69.

12 – Four previously unknown moons

GALILEO GALILEI : 1564 – 1642

Galileo was fascinated by anything that moved. At first it was the pendulum and its regular swing. Then it was objects moving down a slope or falling from a height. He turned his attention to water moving through a siphon and how objects floated. He realised that it needed a force to get something going or, if it was already moving, to stop it, slow it down or turn it from its course. He investigated the path taken by a projectile. When the telescope was invented he discovered moons orbiting the planet Jupiter and ever-changing sunspots carried around a rotating sun. All this was achieved by a man whose father had intended him for a physician.

Galileo was born in or near Pisa, on February 15, 1564. His education and early career swung between Pisa and Florence until 1592. In that year he took the Chair of Mathematics at Padua University. Among his many talents were painting, music, and "explaining, by simple and obvious things, others which are more difficult or abstruse."[1] He was also quite good at upsetting people. They included some who thought Aristotle incapable of being wrong. One upholder of that ancient philosopher wrote, " ... these modern mathematicians [he

CHAPTER TWELVE

probably had Galileo in his sights] solemnly declare that Aristotle's divine mind failed to understand maths, and that as a result he made serious mistakes." Galileo, on his copy of this author's book, noted: "And they are right in saying so for he [Aristotle] committed many and serious blunders, though neither so many nor so silly as does this author every time he opens his mouth on the subject."[2] Galileo further deplores those who " ... attempt to learn from Aristotle that which he neither knew nor could find out, rather than consult our own senses and reason."[3]

Making observations and then thinking carefully about what they meant lay at the heart of Galileo's contribution to science. He was very confident in his own ability, but not without reason. In his last year at Padua he wrote a letter about a series of public lectures he had been giving on his astronomical observations. "The whole university turned out, and I so convinced and satisfied everyone that in the end those very leaders who at first were my sharpest critics and the most stubborn opponents of the things I had written, seeing their case to be desperate and in fact lost, publicly stated that they were not only persuaded but are ready to defend and support my doctrines against any philosopher who dares to attack them ... it was necessary that truth should remain on top."[4]

It is only in this way that science progresses. "Doctrines", observations, theories, proposals can make no headway unless those who put them forward can do so with evidence that convinces and satisfies those who know about such things. In Galileo's day these would have been the best minds of Europe, able to take in the whole sweep of science. Today, the field is so vast that it is more likely to be a relatively small group of international experts that needs to be persuaded.

Four previously unknown moons

CHAPTER TWELVE

In 1610 Galileo left Padua for Florence and became philosopher and mathematician to the Grand Duke of Tuscany. It is difficult for us to get into the minds of those who saw the universe as a series of crystal spheres carrying perfectly spherical heavenly bodies in perfectly circular orbits around a unique centre of rotation – Earth. Everything except Earth, spoilt by the Fall, was thought to be perfect and unchanging. By the late 16th century, the observations and theories of the astronomers were beginning to sow turmoil among these long-held ideas. It was becoming increasingly difficult to reconcile the idea of circular orbits with what was actually seen. A supernova – a new star – observed in 1572, suggested that the heavens were not in fact unchanging. Then, Tyco Brahe's observations on a conspicuous comet in 1577 raised another question. How could a comet break through the crystal spheres surrounding Earth? However, whether or not these spheres really existed, was already controversial. In 1609 came Kepler's proposal of elliptical, instead of circular, orbits. It was on January 7, 1610, using the newly invented telescope, which he himself had significantly improved, that Galileo saw the four previously unknown moons orbiting Jupiter. This meant that Earth was no longer the centre about which **all** heavenly bodies rotated. He went on to describe mountains on the moon, which never again could be thought of as perfectly spherical. Worse still, he described spots on the sun's surface – not only a blemish on its perfection but changing before his eyes – in a supposedly unchanging universe. In such ways does scientific observation contribute to changing our thinking about the universe.

Galileo's earlier writings had been approved and applauded by the Church, but in 1613 his *Letters on Sunspots* showed signs of using the Copernican understanding of the universe as fact rather than theory. Some among the Church authorities detected a conflict with their interpretation of Scripture and in 1616 Galileo was forbidden to hold, teach or defend Copernican astronomy. He agreed to comply but a book he published in 1632[5] was so clearly in opposition to the 1616 edict that sale of the book was banned and its author summoned to Rome. He was sentenced to penance and to house arrest for the remainder of his life.

At first he was confined at the palace in Siena of the Archbishop Piccolomini whose kind indulgence restored Galileo's spirit but aroused the Pope's suspicion. In mid-December 1633, he was returned to the quiet of his own villa in Arcetri, which became his prison. Writing to a friend, he celebrated "two sources of perpetual comfort, first, that in my writings there cannot be found the faintest shadow of irreverence towards the Holy Church; and second, the testimony of my own conscience, which only I and God in Heaven thoroughly know. And he knows that in this cause for which I suffer, though many might have spoken with more learning, none, not even the ancient Fathers, has spoken with more piety or with greater zeal for the Church than I."[6] In *Galileo's Daughter*, Dava Sobel does us the service of throwing a revealing domestic light on this episode.

The Galileo affair is sometimes portrayed as a simple conflict between a rigid church and an enlightened science. It is far more complex, and the personalities and issues involved have inspired whole books on the subject. The period of arrest bore fruit in a last book, *Two New Sciences,* which was smuggled out of Italy and published in Holland. Although blind for the last few years of his life, he kept his mind wonderfully active to the end, which came on January 8, 1642.

When Galileo stated that it was "necessary that truth

CHAPTER TWELVE

should remain on top", he wrote within a scientific context. When science uses the word 'truth'. it refers to our understanding of the universe based on current interpretation of the evidence available. To use Scripture in support of, or against, the findings of science is inappropriate. Galileo expressed concern about "those who, being either unable or unwilling to comprehend the experiences and proofs used in the support of the new doctrine[7] by its author and his followers, nevertheless expect to bring the Scriptures to bear on it. They do not consider that the more they cite these, and the more they insist that they are perfectly clear and admit of no other interpretations than those which they put on them, the more they prejudice the dignity of the Bible ... If they wish to proceed in sincerity, they should by silence confess themselves unable to deal with such matters."[8] We need to reflect carefully on this.

Notes

(1) S. Drake *Discoveries and Opinions of Galileo* Doubleday Anchor 1957 p 265
(2) S. Drake p 223
(3) S. Drake p 143
(4) S. Drake p 60
(5) *Dialogue Concerning the Two Chief World Systems, the Copernican and the Ptolemaic*
(6) S. Dobel *Galileo's Daughter* Fourth Estate 1999 p 329
(7) The reference here is to the Copernican theory but illustrates Galileo's dismay at inappropriate use of Scripture.
(8) S. Drake p 209/210.

13 - God is being celebrated in astronomy

JOHANN KEPLER: 1571 – 1630

Kepler, using his parents' horoscopes, described his father as "criminally inclined, quarrelsome, liable to a bad end" and his mother as "thin, garrulous and bad tempered".[1] If there is any truth in these descriptions the couple seem an unlikely source for a "founder of physical astronomy", but perhaps the implication is that none of us has to be shackled to our origins. We also do well to remember that, although he was intrigued by astrology, Kepler wrote, "If astrologers sometimes do tell the truth, it ought to be attributed to luck." Certainly, he cared enough for his mother to spend time and effort defending her against a charge of witchcraft.

Johann was born on December 27, 1571, in Weil, Germany. He survived smallpox in early childhood, and at age 13 entered Adelberg Seminary. In 1588 he moved to Tübingen University where Michael Maestlin introduced him to the ideas of Copernicus. Six years later he turned, reluctantly, from the

CHAPTER THIRTEEN

priesthood to which he had been heading, to become Mathematician at the Lutheran School in Graz. Here he met the wealthy heiress whom he married on April 27, 1597, "under calamitous skies". She died fourteen years later. I regret not having a sight of an extraordinary surviving document[2] that has Kepler's notes on the eleven prospective brides that appear to have been available to him when he decided to remarry. It seems he felt sure God led him back to number 5. Accordingly, his choice fell on Susanna Reutlinger, a 24-year-old orphan, and seventeen years his junior.

Meanwhile, military and religious upheavals in Europe made for troubled times and on January 1, 1600, Kepler had to leave Graz. He became assistant to the famous Danish astronomer, Tycho Brahe, who by then was working from an observatory at Benatek near Prague. Brahe, without the aid of a telescope, had devised and made measuring instruments with which he was compiling the most accurate astronomical tables then known. When Brahe died in October 1601, Kepler inherited the records on which so much of his own work was based. Referring to Brahe, he wrote, " ... I build this entire structure from the bottom up upon his work, all the materials being borrowed from him."[3]

What he built included a group of three striking and original proposals about how planets moved, the first being 'that planets move in ellipses'. Until then, all attempts at explaining planetary motion had been based on the circle. The circle was held to be the "perfect" figure, appropriate to a perfect God; therefore the universe *must* be based on spheres and circles. This was the accepted wisdom, but Kepler was prepared to question the accepted wisdom when it clashed with Brahe's observation and his own reasoning. Observation and reason

God is being celebrated in astronomy

are the two foundations on which science is built. Science grows and changes because new observations may lead to new thinking, or because new reasoning may suggest other things to look for. Kepler's proposals came to be known as "laws" because, so far, no new observations or reasoning have caused astronomers to change their minds about them.

At the beginning of the most famous of his many books, *Astronomia Nova (The New Astronomy)* Kepler wrote, "My aim in the present work is chiefly to reform astronomical theory ... so that our computations from the tables correspond to the celestial phenomena."[4] His fame lies in the fact that, in essence, his aim was achieved. This clear aim did not prevent him, in the same book, sharing his thoughts on how Scripture should be interpreted. *Ecclesiastes* 1: 4 reads: "Generations come and generations go, but the earth remains for ever". The Hebrew word 'amad' (remains) has the meaning 'to stand' or 'to stand still' and some had used this verse to argue that the earth was stationary and that therefore the sun must revolve around it, in contradiction to Copernicus. Kepler writes, "Does it seem here as if Solomon wanted to argue with the astronomers? No; rather he wanted to warn men of their own mutability".[5] In other words, the author of *Ecclesiastes* was not writing about astronomy: he was simply emphasising how quickly our lives pass, compared with the age of the Earth.

Kepler saw no contradiction between his science and his faith. He gave this advice to his fellow astronomers: "I implore my reader, when he departs from the temple and enters astronomical studies, not to forget the divine goodness conferred upon men ... I hope that, with me, he will praise and celebrate the Creator's wisdom and greatness which I unfold for him ..."[6] "I wanted to become a theologian," he wrote.

CHAPTER THIRTEEN

"For a long time I was restless. Now, however, behold how through my effort God is being celebrated in astronomy."[7]

Notes

(1) ed. C. Gillispie *Dictionary of Scientific Biography* Charles Scribner's Sons 1981 Vol. 7 p 289
(2) A. Koestler *The Watershed: A Biography of Johannes Kepler* University Press of America 1960 p 227
(3) J. Kepler, tr. W. Donahue *The New Astronomy* Cambridge University Press 1992 p 51
(4) J. Kepler p 48
(5) J. Kepler p 63
(6) J. Kepler p 65
(7) ed. C. Gillispie Vol. 7 p 291.

14 – Anatomy ... from the fabric of nature

WILLIAM HARVEY : 1578 – 1657

The life of William Harvey was interwoven with affairs of state during a tumultuous period of British history. He was the eldest of nine children of his father's second marriage and was born in Folkestone, Kent, on April 1, 1578. After the King's School, Canterbury, his father sent him, in May 1593, to Gonville and Caius College, Cambridge. To obtain a medical degree he moved to the University of Padua, Italy, in 1600, where anatomy was well taught by Fabricus. When William arrived in Padua, thirty-six-year-old Galileo was Professor of Mathematics there. Surely when the young Doctor of Medicine returned to England in 1602 he must have been fired also by Galileo's passion for truth, as revealed by observation and experiment. The skill in dissection that Harvey learned in Italy was applied not only to human bodies, but also to eighty or more different kinds of animal, not least the deer of the royal hunt. "I propose to learn and to teach anatomy not from books but from dissection, not from the positions of philosophers but from the fabric of Nature."[1]

Soon after his return to England he married Elizabeth, daughter of the physician, Lancelot Browne. There were no

CHAPTER FOURTEEN

children and we know more about Elizabeth's pet parrot than we do about Elizabeth herself. "My wife had an excellent and a well instructed Parrat, which was long her delight ... the Parrat, which had lived many years sound and healthy, grew sick and being much oppressed by many convulsive motions, did at length deposite his much lamented spirit in his Mistresses bosom, where he had so often sported." It seems that the Harvey scalpel spared nothing, for he continues, "When dissecting his carkase ... "[2]

He received his licence to practise from the College of Physicians in October 1604, and remained very much involved

with the college all his life. He was appointed "Phisicion of St. Bartholomewes Hospitall" on October 14, 1609. At least once a week he was charged to visit the poor in the hospital: " ... You shall not for favour, lucre or gaine, appoynte or write any thing for the poore, but such good and wholsome things as you shall thinke with your best advise will doe the poore good, ... "[3]

Harvey became Physician Extraordinary to James I in 1623 and to Charles I on his accession. From 1630 he was promoted to Royal Physician in Ordinary. In this office he became a

Anatomy ... from the fabric of nature

personal friend of the King and attended him through the Civil War including, at some risk, the battle of Edge Hill. In 1642 his London home was ransacked by Cromwell's troops. " ... some rapacious hand or other not only spoiled me of all my Goods; but also (which I most lament) have bereft me of my Notes, which cost me many years industry."[4] From December of that year he was with the King at Oxford until his Majesty left the City in disguise in April 1646. For his loyalty to the King the Commonwealth fined him £2000, as a 'delinquent'.

In 1647, by then aged nearly sixty-nine, the doctor retired to London to live out the rest of his life, probably a widower, in the homes of one or other of his surviving brothers. A friend who visited during these years recalled that " ... anyone who talked with him in his old age came away not only instructed but also feeling more cheerful."[5] His many references to a Supreme Intelligence suggest something more than the formality of his times. "I do indeed acknowledge God, the supreme and almighty Creator, as being universally present in the production of all animals, and indicated in his works."[6]

By 1616, Harvey's dissections and experiments had led him to the truth about the circulation of the blood, but he hesitated to publish. Eventually, in 1628, his little book on the movement of the heart appeared in print. It is only about fifty pages but is a model of presentation as he marshals the facts gathered from his years of work and sets out his conclusion. "Now at last allow me to put forward my opinion concerning the circulation of the blood and state it formally to all men. ... it must of necessity be concluded that the blood is driven into a round by a circular motion in living creatures, and that it moves perpetually; and that this is the action or function of the heart, which by pulsation it performs."[7]

His conclusion contradicted almost all the accepted

CHAPTER FOURTEEN

knowledge, so that he was very fearful for the reception of his discoveries, " ... so new and unheard of are they, that not only do I fear some mischief to myself from the malice of certain persons, but I am likewise afraid lest I have all men to my enemies ... However, the die is now cast and I put my trust in the love of truth and in the integrity of learned minds."[8]

His fears were partly justified. Referring to his new discovery he wrote, "Some tear the as yet tender infant to bits with their wranglings, as undeserving of birth; others by contrast consider that the offspring ought to be nurtured, and cherish it and protect it by their writings."[9] On the whole, he did not enter into controversy, saying, "Perish my thoughts if they are empty and my experiments if they are wrong ... If I am right, sometime, in the end, the human race will not disdain the truth."[10] And so it turned out.

Harvey's other main work concerned the processes of reproduction and he made very careful studies of developing hens' eggs. He wrote, "But as in the greater world, we say, All things are full of the Deity, so also in the little edifice of a Chicken and all its actions and operations, the Finger of God, or the God of Nature, doth reveal himself."[11]

By all accounts, Harvey was a very active and ever youthful little man, but on June 3, 1657, " ... the little perpetual movement called Dr. Harvye" came to a stop and the "little eie, round, very black, full of spirit" closed.

Anatomy ... from the fabric of nature

Notes

(1) D. Hunter *Harvey and His Contemporaries* Harveian Oration for October 18, 1957   p 19
(See *The Lancet* October 26, 1957   pp 811 - 819)

(2) G. Keynes *The Life of William Harvey* Oxford: Clarendon Press 1966   p 47

(3) J. Da Costa *Harvey and his Discovery* Philadelphia J.B.Lippincott & Co   1879   p 11

(4) G. Keynes   p 162

(5) G. Keynes   p 421

(6) K. Keele *William Harvey*   Nelson   1965   p 56

(7) W. Harvey, tr. G. Whitterlidge *The Movement of the Heart* Blackwell   1976   p 101

(8) W. Harvey   p 74

(9) G. Keynes   p 325

(10) G. Keynes   p 322

(11) G. Keynes   p 426.

15 – Two books

In a famous book called *The Advancement of Learning*, written in 1605, Francis Bacon wrote, "Let no man think or maintain ... that a man can search too far, or be too well studied in the book of God's word, or in the book of God's works; divinity or philosophy:"[1] The "two books" idea has a long history. Raymond Sebonde, a professor at the University of Toulouse who died in 1436, had put it more simply when he suggested that God has given us two books, "The book of nature and the Bible."[2] Reading the "book of nature", "the book of God's works", was what Bacon called 'philosophy' and what we now call science. Later in his book, Bacon refers to "the book of God's word" as "the Scriptures, revealing the Will of God." It is clear that he wants our study of Scripture and Science to go hand in hand.

"Scripture". "Science". What comes to mind when you see or hear those two words? "Boring book – boring subject"? "God and the devil" perhaps? Or could it be "God and God again"?

Scripture – the Bible – has an enormous circulation worldwide. The story of how the Bible came to us in our own language is an exhilarating one, and the same is true for the other 414 languages that have a complete Bible. A further 1068 languages have the complete New Testament and another 873 have a translation of at least one book of the Bible.[3] Ever since seventh-century Caedmon sang "in English verse the

Two books

substance and the themes of scripture",[4] people have worked, and even died, to give us the Bible in our native tongue. Today, English versions proliferate and Bible Societies the world over are putting Scripture into the many languages of people who still do not have a Bible in their mother tongue. The excitement with which such copies are received has to be seen to be believed.

In 1988 a missionary friend wrote from Peru, "Well the great news is that at last the Quechua Bible has been dedicated. ... About 3000 people were at the ceremony and it was exciting to see them literally pushing and shoving after the service to buy their Bible, and then to watch groups crowding around reading it."[5] Why the excitement? Because in their hands they held "the most valuable thing that this world affords."[6] These were the words used by the Archbishop of Canterbury as he presented the Bible to Queen Elizabeth II during her coronation service.

And Science? Can science really have a general appeal? Many people enjoy immensely, and may be profoundly moved by, Music, Literature or Art. These are born of the creativity of women and men; composers, writers, artists; but they themselves are an integral part of this wonderful world in which we find ourselves. No soil – no Schubert. No air – no Austen. No plants – no Picasso. Substitute your own list if you prefer, although I suspect that, for example, the Spice Girls are already proving to have a shorter shelf life than Schubert. Scientists, equally, come from the dust of the ground; no earth – no Einstein. It is the scientist who opens up for us the exquisite intricacies of structure, the marvels of form and function in living things, the elegant laws which govern matter and motion, the unimaginable vastness of this extraordinary universe. No scientific publication is ever going to come anywhere near the

CHAPTER FIFTEEN

worldwide circulation of the Bible, but science can wonderfully enhance both enjoyment and emotion as we look into 'the book of God's works'.

As I sat reading in the garden just now, a small beetle landed upside down on the page. When it righted itself it revealed a highly polished black head and thorax, brilliantly reflecting the sunshine. In contrast, the abdomen was rather sombre but exquisitely marked in a pattern of black and bronze. It rested for a few moments and then lifted the patterned wing cases, unfolded a pair of delicate wings and flew away. I might so easily have brushed the beetle from the page without a thought. Instead, it became a moment of enjoyment. To a beetle expert it would have meant a lot more, and the greater our understanding of science the deeper our appreciation of the world we live in. The same may be said of the Bible. "How many are your works, O Lord! In wisdom you made them all; the earth is full of your creatures."[7]

Bacon urges us to "an endless progress and proficiency in both books"; but not as an end in itself – not just for show or "ostentation". Our studies in God's Word and in God's Works are futile if they simply swell our heads with knowledge and have no outlet in charity, in love, in understanding and in benefit to our neighbour. When he cautions us, " ... do not unwisely mingle or confound these learnings together", he is surely not implying that Scripture and science should not interact. Indeed he writes about one being a key to the other. The word to note is "unwisely" and the world might have been saved a great deal of pain if the Church had heeded it. I shall return to this theme.

Notes

(1) F. Bacon, ed. G. Kitchin *The Advancement of Learning* (1605) Dent 1973 p 8
(2) *Isis* Vol. 47, Part 1, March 1956 p 7 footnote
(3) Bible Society www.biblesociety.org.uk (September 2004)
(4) *The Illustrated Bible Dictionary* IVP 1980 p 446
(5) Rosemary Flack, letter, August 1988
(6) *The Daily Telegraph* 3/6/53
(7) *Psalm* 104: 24.

16 – Touching the spring of the air

ROBERT BOYLE : 1626 – 1691

Science, like faith, captures the hearts of some from all walks of life. Robert Boyle was the seventh son and thirteenth child of Richard, First Earl of Cork and his second wife Katherine. In the father's list of his children, Robert's date of birth is January 25, 1626.[1] The birth took place at Lismore Castle in Ireland. At the age of only eight Robert was sent across to England to be educated at Eton, but from the age of twelve he travelled the European continent for six years with a private tutor, staying some while in Geneva. In Florence, aged fifteen, he read Galileo's *Dialogue Concerning the Two Chief World Systems*. Perhaps it was Galileo's enthusiasm for experiment and observation that inspired the young traveller. Certainly there must have been some spark from these years of travel, which, on his return to England in 1644, ignited the lifelong interest in scientific study.

In 1657 he heard about von Guericke's air pump and, with help from Robert Hooke, improved it and launched into the 43 experiments that made up his first scientific book, published in 1660. The title was *New Experiments physico-mechanical touching the spring of the air and its effects; made for the most part in a new pneumatical engine*. Brevity may be the soul of wit but certainly not of book titles in seventeenth-century science!

Experiment number ten reads as follows: "We took a tallow

candle of such a size, that eight of them made almost a pound; and having a very commodius candlestick let it down into the receiver, and so suspended it, that the flame burnt almost in the middle of the vessel, we did in some two minutes exactly close it up; and upon pumping very nimbly, we found, that, within little more than half a minute after, the flame went out."[2] This was the demonstration that air was necessary for burning and Boyle's written records helped to establish the experimental method as a foundation of science.

To understand what Boyle meant by "the spring of the air" you need only squeeze a balloon or put your thumb over the end of a bicycle pump while trying to push the handle down. It was Boyle who realised that if you double the pressure on a gas you halve its volume (which led him to state the Law that bears his name). In fact, if you press hard enough at low enough temperature, the gas will turn to liquid. Among his experiments was one which showed that sound could not travel through a vacuum.

He also excelled at chemistry and his second scientific book was called *The Sceptical Chemist.* Summing up his *New Experiments* he wrote, "By what has been done, such as it is, there is a way opened, whereby sagacious wits will be assisted to make much farther discoveries in some points of natural philosophy as are yet scarce dreamed of."[3] One of these "sagacious wits" was Newton, one of Boyle's firmest followers.

Boyle never married. He twice spent four-year periods living in Oxford and was a frequent visitor to London. It was there that he settled in 1668, living in the home of one of his sisters. She died on December 23, 1691, and Robert followed her to the grave just a week later. (One account says he lived with her for "the greatest part of 47 years".)

In the opinion of Thomas Birch, Boyle's Christianity was not

CHAPTER SIXTEEN

a nominal affair. "He always considered it as a system of truths, which ought to purify the hearts and govern the lives of those who profess it. ... He loved no narrow thoughts, no low or superstitious opinions in religion; and therefore as he did not shut himself up within a party, so neither did he shut any party out from him."[4] Faith governed his life to the point that, although in 1660 he was a founder member of the Royal Society of London, he declined the Presidency in 1680 because he had scruples about taking an oath. He was certainly the Society's most influential member during the early years.

In his own writing Boyle echoes the "two books" approach to faith. Boyle's version is, " ... whereas there are two main ways to arrive at God's attributes; the contemplation of his works, and the study of his word; I think it may be doubted, whether either or both of these will suffice to acquaint us with all his perfections."[5] He held a very high view of his God. "We ought whenever we speak of God and of his attributes to stand in great awe ... how unable, as well as unworthy, we are to penetrate the recesses of that inscrutable, as well as adorable nature; and how much better it would become us, when we speak of objects so much above us, to imitate the just humility of that inspired poet that said, 'Such knowledge is too wonderful for me; it is high, I cannot attain unto it.'[6]."[7]

Notes

(1) *The Life of the Hon Robert Boyle* p xi in *The Works* (See (2) below)
(2) R. Boyle, ed. T. Birch *The Works* Georg Olms Hildesheim 1965 Vol. 1 p 26
(3) R. Boyle p 117
(4) R. Boyle p cxli
(5) R. Boyle Vol. 5 p 131
(6) *Psalm* 139: 6 (KJV)
(7) R. Boyle Vol. 5 p 157.

17 – Filled with wonder and delight

JOHN RAY : 1627 – 1705

Walking to my car one evening, I noticed, poking through the park railings, some vigorous shoots of an attractive grey-leafed shrub. "Somebody is going to have to clip these back before long," I said to myself, and by the time it came about I had three healthy little cuttings set in pots. I then discovered that something similar had happened in Oxford very much earlier. John Ray had written, " ... we gathered a sprig of it (a sage-like plant) ... in an Oxford garden, put it in a wallet and carried it for three days in a bag: when we got back to Cambridge we planted it: it rapidly recovered and struck root."[1] As I tried to find out more about Ray, I had to show my reader's ticket to get into the Bodleian Library where Ray himself was "entered as a reader" well over three hundred years ago. I find myself taking an exaggerated pleasure in these two snippets of information about someone I have come to admire so much.

John Ray was baptised in the Parish Church of Black Notley in Essex on December 6, 1627, the son of the blacksmith and his wife, " ... a very religious and good woman and of great use in her neighbourhood, particularly to her neighbours that were lame or sick, among whom she did great good, ... "[2]

He attended Braintree Grammar School and it seems likely that a bequest in the will of Thomas Hobbes, and the influence

Filled with wonder and delight

of Samuel Collins, Vicar of Braintree, combined to take him on to Cambridge University in 1644. Here he studied for the priesthood. In 1649, he became a Fellow of Trinity College. He developed an eloquent but down-to-earth style of sermons that put him much in demand, although he was not ordained until 1660.

About nine years earlier, something happened which inspired the other great passion of his life. He had been ill, and, by way of convalescence, had taken to riding and walking in the countryside around Cambridge. The enforced leisure of these journeys opened his eyes to the variety and beauty of plants which, until then, he had trampled underfoot without a thought. "First," he wrote, "I was fascinated and absorbed by the rich spectacle of the meadows in springtime; then I was filled with wonder and delight by the marvellous shape, colour and structure of the individual plants. While my eyes feasted on these sights, my mind too was stimulated. I became inspired with a passion for Botany, and I conceived a burning desire to become proficient in that study."[3]

To his astonishment, he found not a single person in the University who could help him, so he became his own teacher. For the next six years he made himself familiar with the plants in and around Cambridge, recording their habitats with a precision that makes many of the places identifiable even today. A further three years were spent composing, elaborating, revising, " ... and so at last with God's blessing I completed my task, and sent the book to the printer."[4] The title and the book itself were in Latin, which Ray wrote with masterly elegance. It was the first of the many regional floras which have given both help and pleasure to generations of botanists since.

In 1662 he refused, on principle, to subscribe to the Act of

CHAPTER SEVENTEEN

Uniformity and thereby forfeited his Trinity College Fellowship and his right to serve as a priest, a fearful price to pay for integrity. He returned to Black Notley unemployed and with no income and wrote to a friend, "I shall now cast myself upon Providence and good friends."[5]

He found himself in safe hands. With some of these friends, he left Dover in 1663, to travel widely on the Continent, expanding both his botanical knowledge and circle of scientific contacts. One of the friends was Francis Willughby, a former pupil, and together they planned a comprehensive survey of both plants and animals. But Willughby died young. Over the next thirty years Ray produced, in his beautifully clear hand, a whole series of manuscripts; the books on birds and beasts, fishes and insects, plant classification and distribution, on which his fame rests. Some of them used the material left by Willughby and are surely written in tribute to their friendship and joint enterprise. In 1667, Ray was elected to the recently formed Royal Society.

The generosity of friends provided him with successive homes until 1678, when his mother died, and he moved into the house he had built for her in Black Notley. Five years earlier, aged forty-five, he had married the young Margaret Oakley. It was another six years before she gave birth to their first children, twin girls. Two more daughters followed. How one would like to know more of his feelings about this lively family springing up around him at an age when in other circumstances he might have expected to be a grandfather.

In his last years, John must have wished his mother was still around as, racked with almost constant pain from leg ulcers, he struggled, unsuccessfully as it turned out, to finish the book on insects. No longer able to get about himself, his notes record the many catches made for him by the four girls.

Filled with wonder and delight

Jane, the youngest, made her first capture at four and a half. His wife and a Thomas Simson were also among the hunters and one has the impression of a household buzzing happily with activity. But there was one great sadness. On February 1, 1698, he wrote to his friend, the physician Hans Sloane, "My dear child, for whom I begged your advice, within a day after it was received became delerious and at the end of three days died apoplectic ... "[6] It was Mary, one of the twins.

On January 7, 1705, John Ray wrote his last letter to Hans Sloane: "Dear Sir, the best of friends – these are to take a final leave of you as to this world. I look upon myself as a dying man. God requite your kindness expressed anyways towards me an hundredfold, bless you with a confluence of all good things in this world, and eternal life and happiness hereafter, and grant us a happy meeting in Heaven. I am Sir, Eternally yours, John Ray."[7] He died ten days later.

Ray's scientific works in Latin were read in other countries too, but his most popular book was written in English; it went through many editions in his lifetime and well into the century beyond. This was *The Wisdom of God Manifested in the Works of Creation*. In it he wrote, "For being not permitted to serve the Church with my Tongue in Preaching, I know not but it may be my Duty to serve it with my Hand by Writing."[8] Thirty years earlier, in his first book, he had written, " ... I know of no occupation which is more worthy or more delightful for a free man than to contemplate the beauteous works of Nature and to honour the infinite wisdom and goodness of God the Creator."[9]

CHAPTER SEVENTEEN

Notes

(1) C. Raven *John Ray: Naturalist* Cambridge University Press 1986 (Reissue of 2nd ed., 1950) p 109
(2) C. Raven p 9
(3) C. Raven p 22
(4) C. Raven p 24
(5) ed. R. Gunther *Further Correspondence of John Ray* London. Ray Society 1928 p 25
(6) C. Raven p 295
(7) ed. E. Lankester *The Correspondence of John Ray* London. Ray Society 1848 p 459
(8) J. Ray *The Wisdom of God Manifested in the Works of Creation* London 1704 Preface
(9) J. Ray, tr./ed. A. Ewen & C. Prime *Flora of Cambridgeshire* Wheldon & Wesley 1975 p 26.

18 – An astonishing genius

ISAAC NEWTON : 1642 – 1727

Isaac Newton's life began, probably prematurely, on Christmas Day, 1642, and for some time his tiny frame hovered precariously between life and death. Life triumphed and Newton's legacy to science became possible. One of the striking differences between science and the arts lies in the significance of the individual. Had that fragile life perished in 1642, the science we associate with Newton would have come to us by other names in other places at other times. But if, for example, in 1628, Bunyan had succumbed at birth, then the world never would have had "Pilgrim's Progress".

    Isaac never knew his father; he had died during the autumn of 1642. Young Newton grew up in Woolsthorpe, Lincolnshire, where the family home survives. For the first three years he was brought up by his widowed mother and then, when she remarried and moved to North Witham, by his grandmother. In 1653 his mother was again widowed and returned to Woolsthorpe with a stepbrother and two stepsisters for Isaac. In spite of these upheavals, mother and son enjoyed a lifelong affection.

    After schooling in local villages and at Grantham he entered Trinity College, Cambridge, in 1661. Within five years he was producing original mathematics "that would have left the

CHAPTER EIGHTEEN

mathematicians of Europe breathless in admiration, envy and awe"[1] – had they known of his existence. In 1669 his professor wrote to a friend, " ... I am glad my friend's paper giveth you so much satisfaction. His name is Mr Newton; a fellow of our College, and very young ... but of an astonishing genius and proficiency in these things."[2] Even so, the *Principia*, his famous book, which was the eventual fruit of this early burst of mathematical energy, did not appear for nearly two decades and then largely through the efforts of the astronomer Edmund Halley, who saw it through the press at his own expense. On July 5, 1687, Halley wrote: "Honoured Sir. I have at length brought your book to an end and hope it will please you."[3] It pleased the whole world – immensely, though not at once.

We can no more separate Newton and gravity than Morecambe and Wise. Apples don't just fall to the ground, they are responding to the force we call gravity. It was Newton's genius to realise that this force acted between all bodies everywhere in the universe. "I began to think of gravity extending to ye orb of the moon." It did not stop there. He went on to calculate that this mysterious force attracts any two bodies with a strength that depends on the amount of matter in each body and how far they are apart.

What most of us have forgotten, if we ever knew, is that as the apple is attracted to Earth so Earth is attracted to the apple. Life as we know it is based on a particular value for gravity. If it were much stronger autumn leaves might fall like lumps of lead; much weaker and a highjumper might leap into orbit. We owe to Newton not only the law of universal gravitation, but also the laws of motion, and a form of calculus. There were wonderful experiments too, on the nature of light. "No other investigation of the 17th century better reveals the

An astonishing genius

power of experimental enquiry animated by a powerful imagination and controlled by rigorous logic."[4] That sentence contains a good definition of science.

After his academic life in Cambridge, Newton was appointed Warden and eventually Master of the Royal Mint. He was President of the Royal Society for 23 years and was knighted by Queen Anne in 1705. The huge collection of Newton manuscripts still keeps scholars busy, analysing not only his science but also his complex character. He insisted that the two great commandments, to love God and one's neighbour, should be the duty of all nations; but that did not prevent vitriolic relationships with Hooke, Leibnitz and Flamsteed. If some accounts are to be believed, his dealings with Flamsteed were disreputable to say the least. Confessions of faith are always easier than demonstrations of love. In others he did inspire affection and it seems that his generosity

CHAPTER EIGHTEEN

went far beyond the bounds of family.

His physics and maths represent only a portion of his wide-ranging studies, for he read and wrote extensively on alchemy, Scripture and theology, and the early Church Fathers. The English philosopher, John Locke, in a letter to his cousin, wrote: "Mr Newton is really a very valuable man, not only for his wonderful skill in mathematics, but in divinity too and his great knowledge in the Scripture, wherein I know few his equals".[5] With so vast an output of writing, one quotation can hardly sum up Newton's theology, but the following is a striking statement of faith: "We must believe that there is one God or supreme Monarch that we may fear and obey him and keep his laws and give him honour and glory. We must believe that he is the father of whom are all things and that he loves his people as his children that they may mutually love him and obey him as their father ... We must believe that he is the God of the Jews who created the heaven and earth (and) all things therein."[6]

Notes

(1) R. Westfall *The Life of Isaac Newton* Cambridge University Press 1993 p 45
(2) ed. H. Turnbull *Correspondence of Isaac Newton* Cambridge University Press 1959 Vol. 1 p 14
(3) H. Turnbull 1960 Vol. 2 p 481
(4) R. Westfall p 56
(5) R. Westfall p 236
(6) R. Westfall p 303.

19 – Looking stars in the face

JOHN FLAMSTEED : 1646 – 1719

John Flamsteed was born in Denby, some seven miles north of Derby on August 19, 1646, the only son of a Derby businessman. His mother died when he was three and his stepmother only five years later, each having left him a sister. At fourteen he contracted an illness which left him almost too weak to go to school, but neither physicians nor a visit to Ireland to seek faith healing were any help to him in the short term. He recorded that he was nineteen years, six days and eleven hours old when he left for Ireland, indicative perhaps of his precise approach to recording throughout his life.

It seems his illness may have provided the opportunity to read about astronomy, for by the time he was twenty-three he was able to write to the President of the recently-formed Royal Society offering his "calculations of such of the more notable phenomena of the year 1670, as will be conspicuous in the English horizon, if the heavens be clear ... I hope you will not account me culpable for having adapted the calculations to the meridian of a place no more famous than Derby."[1] He makes the point that such a meridian is nearer the centre of England than that of London. It is interesting to consider how close we came to having Derby Mean Time instead of GMT. He did not

CHAPTER NINETEEN

know then that in little more than five years he would be appointed "astronomical observator" by Charles II, in effect the first astronomer royal. The still-surviving Greenwich Observatory was completed for him within the year.

The progress of science depends in part on the patient, painstaking, persevering collector of data. John Flamsteed fits this description. We have already seen how vital to Kepler were the observations of Tycho Brahe. However, Flamsteed found inaccuracies in these observations. He wrote to his patron, Sir Jonas Moore, on July 16, 1678, "Ever since I perceived the fault of the Tychonic Catalogue I have determined not to rely on any part of it, but if God bless me with health and success in my endeavours, to begin a new one ... "[2] None the less, he was full of praise for what Brahe had achieved.

Four years earlier Moore had written concerning someone else, that it troubled him "that anybody should busy themselves about the stars and never look them in the face."[3] Flamsteed spent the whole of his lifetime "looking stars in the face" and meticulously recording his tables of observations. In August 1716, he wrote to a friend, Abraham Sharp: "I find my strength impairs daily, so that I can now walk but once a day to Church on Sundays. My memory and reason continue still; so as that I have but little cause to complain. I bless God for it."[4] Three years later, with his monumental catalogue of observations unfinished, he was taken ill just after Christmas, and died as the year itself drew to a close.

But for the labours of Joseph Crosthwaite, his loyal assistant, and Abraham Sharp, the three volumes of his *Historiae Coelestis Britannicae* might never have seen the light of day. Crosthwaite wrote, "The love, honour and esteem I have, and shall always, for his memory and everything that belongs to him, will not permit me to leave Greenwich or London, before I hope the three volumes are finished."[5] Sad to record, he received "not one farthing" from Flamsteed's widow "for all my time spent and all my expenses in attending the printing and map."[6]

In 1675, the year of his appointment to Greenwich, Flamsteed had been ordained deacon in the Anglican Church. In 1684 he accepted the living of Burstow in Surrey. His letters reveal evidence of several visits to Burstow, but the job of ensuring that his parishioners did not suffer from his dual responsibilities presumably fell to a curate.

At Greenwich, his making of observations suffered many interruptions. He had sent to him "two bluecoate boyes – to learne the stars. These I am forced to instruct daily." In summer he complained that "many of my friends and acquaintances and sometimes crowds of strangers come to visit this place." It is not difficult to imagine what would be his feelings towards the crowds of strangers, tourists and school parties mostly, that today swarm noisily through the observatory and his private quarters. There was illness too: "I have almost constantly some aches and sometimes pungent pains."[7] Faulty instruments were yet another hazard. A wall quadrant sounds harmless enough but he records that the one made for him by Hooke "has often deceived me ... I tore my hands by it and had like to have deprived Cuthbert [his assistant] of his fingers."[8]

Flamsteed had known Newton well, and at first cordially, as he supplied him with relevant observations. However, he was furious when, in 1712, Newton and Halley published some of his results prematurely and against his will and, adding insult to injury, "with corrections". He was eventually able to retrieve three-quarters of this edition, some three hundred copies,

CHAPTER NINETEEN

which he gladly "sacrificed to heavenly truth" on his bonfire. Flamsteed, in his dual role of scientist and clergyman, demonstrated a concern for truth and accuracy, which both science and faith neglect at their peril.

Notes

(1) ed. E. Forbes et al *Correspondence of John Flamsteed* Institute of Physics Publishing 1995 Vol. 1 p 12
(2) ed. E. Forbes p 643
(3) ed. E. Forbes p 291
(4) F. Bayly *An Account of the Reverend John Flamsteed* Dawsons 1966 p 324
(5) F. Bayly p 333
(6) F. Bayly p 363
(7) ed. E. Forbes p 638
(8) ed. E. Forbes p 645.

20 – The naming of plants and animals

LINNAEUS (CARL VON LINNE) : 1707 – 1778

"Now the Lord God had formed out of the ground all the beasts of the field and all the birds of the air. He brought them to the man to see what he would name them; and whatever the man called each living creature, that was its name."[1]

In the Genesis account there is no reference to the naming of plants. Nevertheless, several named plants feature in the Bible although we cannot always be certain precisely what they were. Elijah, fleeing from Jezebel, settled under "a broom tree" and prayed that he might die.[2] We are no more certain of the tree's real identity than we are of the "vine" under which Jonah sulked.[3]

Botanists now have books called 'Floras' to identify the plants found within a given area, such as a nation or a county. One of my own early pleasures was using the short 'Flora' of the day to find the names of plants I came across as weeds in the garden, or on country walks. Although using such books can be a frustrating business, success is sweet and I well remember my first identification. It was one of the little blue

CHAPTER TWENTY

speedwells found in gardens and fields. Its modern name is 'common field-speedwell', although it was then known as 'Buxbaum's speedwell'. So far, I have not found out who Buxbaum was, but his speedwell was only introduced into Britain, from Western Asia, in 1825. Some alien plants only establish a precarious bridgehead. This speedwell has achieved a wholesale invasion.

Most of the plants and creatures we see in the countryside have common names, especially if they are showy or have had some use to us for medicine or in other ways. For example, the foxglove, with the Latin name *Digitalis purpurea,* is not only showy but yields the drug digitalin, still useful in treating heart complaints. Sometimes plant names differ in different parts of a country and, of course, from country to country. Plants are no respecters of county or national boundaries. In the days when Latin was the common language of an educated élite, plants and animals were burdened with long, descriptive names, but at least the name would be recognised internationally. For example, the lesser periwinkle, a plant of woods and shady hedgerows, was *Clematis Daphnoides minor, seu Vinca Pervinca minor.*

The man who brought a much-needed brevity to the naming of plants and animals was the Swedish naturalist, Carl von Linné. He took this name when he was ennobled in 1757 but is better known in the English-speaking world as Linnaeus. He gave each animal or plant two names only: one for the group, or Genus and one for each particular type, or Species. Wherever it grows, the common field-speedwell is now known as *Veronica* (the genus) *persica* (the species: the name that separates it from the other species in the same genus). It was Linnaeus who named us *Homo sapiens.* In the British Isles, it would now be extremely rare to find even a moss or fungus or beetle, let

The naming of plants and animals

alone a flowering plant or mammal, which has not been carefully described, and given its two names. This is not the case world-wide. Recently a new species of deer was discovered, in the forests of Vietnam – a very unusual event. Sadly, human activity, such as agriculture, development and tourism, means that many thousands of less conspicuous animals and plants are disappearing before science has been able to name and describe them. Peter Crane, Director of Kew Gardens, is reported as saying, "It's clear that a huge extinction is going on around the world. We are losing plants we don't even know we have."[4]

Carl was the first of five children in the family of a Lutheran curate, Nils, and his wife Christina. Nils had a deep enthusiasm for his garden and for the wild plants nearby. Carl took after his father in this, and considered the shore of Lake Mockeln near his birthplace "one of the most beautiful places in all Sweden ... When one sits there in the summer and listens to the cuckoo and the song of all the other birds, the chirping and humming of the insects; when one looks at the shining, gaily coloured flowers; one is completely stunned by the resourcefulness of the Creator."[5]

At age 9 he attended the grammar school, thirty miles from home, where his education prepared him for university, first at Lund and a year later at Uppsala. Both parents hoped that Carl, too, would become a pastor, and in the summer vacation of 1727 his mother "still clung with pathetic obstinacy to a hope that a year at Lund might have turned Carl's thoughts to ordination; however, when she saw he did nothing but glue plants to paper she had to admit that the battle was lost."[6] It is interesting to note that many scientists *were* ordained clergy, while others, such as Kepler and Darwin had been heading in that direction. There are those among today's clergy who

CHAPTER TWENTY

began life as scientists.

As a young man, Linnaeus travelled quite widely, and visited England in 1736. Later, he left the travelling to his many "disciples" and waited eagerly for their return, laden with their botanical finds. By 1742, when he became Professor of Botany at Uppsala University, he had already been questioning existing systems of classification for more than a decade and had begun to publish his own ideas, for example in 1735 in the book known as *Systema naturae.* The number and arrangement of the stamens and pistils, the sexual parts of flowers, had caught his eye, and this became the basis of his classification. It did little to place plants in natural groupings but had the merit of convenience. His "Binomial [two names] System", put forward in 1753, is of more lasting importance.

A prolific author, famous and honoured at home and abroad, Linnaeus was perhaps more valued than loved. He died January 10, 1778, and was buried in the cathedral at Uppsala.

Linnaeus was bowled over by what he saw around him in "The Book of God's Works". In his introduction to later editions of *Systema naturae* he wrote, "I tracked His footsteps over nature's fields and found in each one, even in those I could scarcely make out, an endless wisdom and power, an unsearchable perfection."[7] He was equally at home in "The Book of God's Word" and quoted freely from it to show his wide-eyed wonder at God the Creator. He may never have taken a further step to enjoy the love of God the Father and His gift of new life in Christ.[8] However, he was a regular if somewhat eccentric churchgoer, as the following story reveals. It seems he considered the length of sermon more important than the content, for he would leave the church after an hour, whether the preacher had finished or not. His dog Pompe always

The naming of plants and animals

accompanied his master to church and, if Linnaeus was indisposed, even went alone, and settled quietly in the family pew. On such occasions Pompe remained faithful to his absent master, and after an hour, would leave the pew and pad silently from the church.[9]

Notes

(1) *Genesis* 2: 19
(2) *1 Kings* 19: 4
(3) *Jonah* 4: 9
(4) *The Times* 25/8/99
(5) W. Blunt   *The Compleat Naturalist*   Collins   1971   p 15
(6) W. Blunt   p 25
(7) ed. T. Frangsmyr   *Linnaeus : The Man and his Work*
    Science History Publications USA   1994   p 12
(8) *Romans* 6
(9) W.Blunt   p 179.

21 – Researches into different "airs"

JOSEPH PRIESTLEY : 1733 – 1804

Mr Keighley of Heckmondwike was an ardent persecutor of eighteenth-century dissenters. His enthusiasm for this activity led him to hide in one of their chapels during a service, in the hope of securing evidence against them. The outcome no doubt surprised him greatly, for he was soundly converted. The event resulted in Joseph being brought up in a dissenting household, for Mrs Keighley was his aunt and she adopted Joseph when his mother died.

Joseph later demonstrated the independent thinking that marked his life when his views were found to be unorthodox enough to stop him becoming a communicant member of the Chapel. However, he found his manner of thinking was encouraged by the Dissenting Academy at Daventry. "We were permitted to ask any questions, and to make whatever remarks we pleased; and we did it with the greatest, but without an offensive freedom."[1] Until 1871 the universities of Oxford and Cambridge were barred to dissenters, but various nonconformist academies had provided a remarkably effective alternative. The one at Daventry was successor to the famous Northampton Academy established by the hymn-writer, Philip Dodderidge.

Researches into different "airs"

In 1755, aged 22, Joseph accepted his first assistant pastorate at a chapel in Needham Market. It was an unhappy start but he fared better at Nantwich where he also started a school, and made enough money to buy a few books and some scientific equipment, and thus indulge his passion for experimenting.

His versatility then took him to Warrington as a language teacher. While there, in 1761, he married Mary Wilkinson and in 1766 he was elected to the Royal Society on the strength of his experiments in electricity. From 1767 he combined a nonconformist ministry at Mill Hill Chapel, Leeds, with researches into different "airs" or gases. Perhaps it was not by chance that he lived next to a brewery, for brewing produced plenty of carbon dioxide for his experiments.

From 1773, Priestley served the Earl of Shelburne as librarian and literary companion at Bowood House near Calne in Wiltshire. He was delighted to be able to continue his experiments. The house remains in the same family today and is open to the public so that, if you wish, you may visit the very room which he used as a laboratory. There, with his new burning glass he tried heating many different substances to see if they gave off gases. On 1st August 1774 it was the red oxide of mercury on which he focused the sun's rays. He had earlier invented a way of collecting the gas given off – perhaps you remember the pneumatic trough from your school days. On this occasion he tested the gas with a lighted candle. He reported the result in *Experiments and Observations on Different Kinds of Airs* and wrote, "But what surprised me more than I can well express, was, that a candle burned in this air with a remarkably vigorous flame." He had discovered oxygen. We now know that Carl Scheele, in Sweden, had done so in 1771, but it was Priestley's account that first became public

CHAPTER TWENTY-ONE

knowledge.

Later, in 1774, he went with the Earl on a continental tour, during which he met Lavoisier, who held a very high regard for Priestley's experiments. Commenting on the latest ones submitted to the Royal Society in London, Lavoisier wrote, "This may be regarded as the most painstaking and interesting work dealing with the fixation and liberation of air [gas] which has appeared since that of Mr. Hales."[2] The Rev. Stephen Hales (1677 - 1761), of Teddington, had been an ingenious inventor and experimenter, especially in collecting and studying the gases exchanged between plants and the air.

After leaving Lord Shelburne in 1780, Priestley became a minister at New Meeting in Birmingham. Here, in the Lunar Society, he found congenial companions in Erasmus Darwin – Charles' grandfather, Josiah Wedgwood – the potter, James Watt – of steam-engine fame, and others.

However, his pamphleteering against government policy in the American colonies, his sympathy with the French Revolution and no doubt his nonconformity also made him enemies. They may have been responsible for rumours circulating in Birmingham which, on July 14, 1791, inflamed a mob to burn down the chapel and loot his house and laboratory. He escaped with his life and he and his wife lived apprehensively in London for three years before emigrating to America in April 1794. Their sons were already in America and they settled in Northumberland in Pennsylvania to be near them. Sadly, Mary died in 1796 before the house they were building was completed. Joseph survived her by eight years, still experimenting and writing, until his death on February 6, 1804.

Mrs Priestley must often have wished that her husband would keep his mouth and his inkwell closed. But he thought deeply and would not be silenced even though it cost him his

Researches into different "airs"

home, his beloved chapel and laboratory and, in the end, his country. It was said of his theology that he "had too little Christianity for fervent believers and too much for complete infidels."[3] His son wrote, " ... though my father by the fearless avowal of his opinions created many enemies, yet the honesty and independence of his conduct procured him many friends."[4]

For himself, he never doubted "the guiding and sustaining hand of God". It was Providence he thanked for his scientific pursuits which, "next to theological studies, interest me most. Indeed, there is a natural alliance between them, as there must be between the words and works of God."[5] On the day before he died, it was of resurrection that Priestley asked his son to read. The eleventh chapter of John's gospel concerns the raising of Lazarus. "Jesus said to her [Martha], 'Your brother will rise again.' Martha answered, 'I know he will rise again in the resurrection at the last day.' Jesus said to her, 'I am the resurrection and the life. He who believes in me will live, even though he dies; and whoever lives and believes in me will never die'."[6] Priestley, in a long life of reading and studying the Bible had come to accept the truth and authority of the words of Jesus recorded there. In the hour of his death they were a powerful assurance of what lay ahead for him.

Notes

(1) W. Aykroyd   *Three Philosophers*   Heinemann   1935   p 36
(2) W. Aykroyd   p 63
(3) W. Aykroyd   p 62
(4) Priestley & Son   *Memoirs of Dr Joseph Priestley*   Vol. 1   p 205
(5) W. Aykroyd   p 213
(6) *John* 11: 23 – 25.

22 – The Book of God's Word

On April 26, 1997, television showed pictures of a dugout canoe laden with exhausted refugees from war-torn Zaire, as it then was. It seemed that they had nothing at all, other than the clothes they wore. But there was one exception: a young man who held in his hand a Bible. If you could only salvage one item among all your possessions, would it be a Bible?

When I was a child, I was encouraged to read that wonderful book for myself. I was fortunate in having parents I responded to out of love rather than fear, and I owe them a great debt for the habit formed in childhood of reading daily in the "Book of God's Word". Early in 1963, my father became aware that he was suffering from the wasting disease which took his life two years later. In June of that year he wrote, "Truly I can say life has been very wonderful and blessed for me these past two or three months – they have held some of the happiest days of my life. I have known a blessedness and content in being in this wonderful world – everything has spoken to me of the work of our Creator. The great rhythm of life goes on and we must take our part in it – seed time and harvest are natural things and I shall be part of the history of our beloved land whose countryside has been so much deep joy to me ... My favourite readings are Psalm 23 and the opening of John 14. I have been rising early to sit in the sunshine reading – I have put some slips of blotting paper where I have been reading – what comfortable words and true."

The book of God's Word

My father was only one among multitudes that have mined these rich veins of comfort down the centuries but I am glad I still have that Bible with the slips of paper where he placed them. Rich source of comfort though it is, I think the Bible is even more important in revealing the nature of God; his holiness, his justice and his overwhelming love for us. "Know therefore that the Lord your God is God; he is the faithful God, keeping his covenant of love to a thousand generations of those who love him and keep his commands."[1]

The nature of God is seen supremely in the life of Jesus, his death for us, and his resurrection. When John wrote his gospel he made it clear that what he included about Jesus was written for a specific purpose – " ... that you may believe that Jesus is the Christ, the Son of God, and that by believing you may have life in his name."[2] God is faithful, but it takes two to covenant.

The Bible can be read at several levels. For sheer beauty of expression there are passages from the Authorised Version of 1611 which will surely remain matchless for as long as language lasts. The Bible is also read as a subject for study, perhaps to unravel the history of the people of Israel or of the early Church. It may be subjected to textual analysis or an investigation as to how it came to be, or how it has survived to our day. Many, however, believe there is far more to the Bible than this: that it is the living word of God and that through it we may hear him speaking to us if we approach it with an open and receptive spirit.

As we have seen, on the day before he died, the chemist, Joseph Priestley, asked his son to read from the 11th chapter of John. "I was going to read to the end of the chapter," wrote the son, "but he stopped me at the 45th verse. He dwelt for some time on the advantage he had received from reading the scriptures daily, and advised me to do the same ... "[3] In his

CHAPTER TWENTY-TWO

Memoirs, Priestley had earlier written, "So strongly is my mind impressed with a sense of the importance of the habitual reading of the scriptures ... that I do not see how those persons who neglect it, and who have no satisfaction in habitually meditating on the infinitely important subjects to which they relate, can be said to have anything of Christianity besides the name."[4]

When Paul sent his first letter to the young church at Thessalonica in Greece he wrote: " ... when you received the word of God, which you heard from us, you accepted it not as the word of men, but as it actually is, the word of God, which is at work in you who believe."[5] Of course the Bible as we know it did not then exist but as we have seen, it has for many centuries been referred to as "the Book of God's Word". Certainly it has been the experience of Christians through these centuries that, given the opportunity, the Bible is "at work in those who believe". A Bible on the shelf has no chance. It has to be in the hand, in the head, and eventually in the heart if it is to do its life-changing work.

Notes

(1) *Deuteronomy* 7: 9
(2) *John* 20: 21
(3) Priestley & Son   *Memoirs of Dr Joseph Priestley*   Vol. 1   p 217
(4) Priestley & Son   Vol. 2   p 21
(5) 1*Thessalonians* 2: 13.

23 – The true nature of burning

ANTOINE LAVOISIER : 1743 – 1794

I remember, in my school laboratory, holding a strip of the silvery metal, magnesium, in a Bunsen burner flame, and being amazed to see it burst into dazzling fire. Another piece of the metal, carefully weighed beforehand, was then burnt (calcined) in a little porcelain crucible and the white powder left behind was, just as carefully, weighed again. I was less excited to find that this white ash, or calx, weighed *more* than the original metal. But how extraordinary this result is. The common sense view is that things become lighter when they burn. Which just goes to show that something more than common sense is needed for progress in science.

In the seventeenth century it was believed that when something burned it released a mysterious substance into the air. In 1697 the German chemist Georg Stahl named the mystery matter, 'phlogiston'. How surprising then was the demonstration by Guyton in 1771 that metals *increased* in weight when burnt in air. How could this be, if phlogiston was released? Could phlogiston weigh less than nothing?

It was Lavoisier's genius, demonstrated by a series of experiments and some clear thinking, that revealed the true nature of burning. As is usually the case, he did not work in isolation. In 1627 Stephen Hales had produced his book

CHAPTER TWENTY-THREE

Vegetable Staticks which included experiments on air. Lavoisier wrote of this book, "The Author has everywhere joined to them [his experiments] views entirely new ... Too much cannot be said to induce the reader to peruse that Author's own work. He will find a most inexhaustible fund of meditation."[1] He paid similar tribute to Joseph Priestley. In 1773 Lavoisier read all he could find concerning burning and by 1778, after painstaking experiments, he was able to write that the substance which unites with metals when they burn is "nothing more than the healthiest and purest part of air", that which he later named 'oxygen'. Further and dramatically, he could say, "I have established that the air of the atmosphere is not a simple substance, an element, as the Ancients believed and as has been supposed to our own time."[2] A few years later he was able to say the same of water. For two thousand years the four elements, earth, air, fire and water had been the basis of thinking about the world. Now that mode of thinking would no longer pass muster.

On May 8, 1794, another crop of severed heads fell to the blade of the guillotine in La Place de la Révolution in Paris. In that hideous harvest was the head of Antoine Lavoisier. That morning, along with his father-in-law and 26 other associates of the Ferme Générale, he had been condemned to death by the Revolutionary Tribunal. The Ferme Générale, by then disbanded, had been a consortium collecting certain taxes on behalf of the government. These unfortunate men paid dearly for their involvement with it.

Writing on a spring day just over 200 years later, something of the sheer horror of these summary executions fills my mind. It seems incredible that a scientist, already so famous, should suffer such a fate. In England he might have expected burial in Westminster Abbey. As it was his corpse was tipped into an

The true nature of burning

unmarked trench, just one more of the many, many victims of the Terror.

Antoine had been born in Paris just over fifty years earlier. Both his parents came from families of lawyers, but his mother died when he was five and he was reared by an adoring aunt. The youth, too, trained for the law but, inspired by geology and chemistry teachers, his interest turned to science. His ability and ambition and perhaps a bit of string-pulling saw him elected to the French Academy of Science in March 1768.

Three years later, in circumstances which would not look out of place as the plot for an opera, he married the fourteen-year-old daughter of one of his seniors in the Ferme Générale. She proved a wonderful collaborator in her husband's scientific work, writing notes and drawing apparatus as he delved into the mysteries of vaporisation and heat, the formation of acids, the composition of water as well as the experiments on combustion and the properties of "aeriform fluids". Thanks to Lavoisier, the last mentioned became known as "gases" in a thorough modernisation of chemical language. Remarkably, all this work was done between the hours 6.00 - 8.00 a.m. and 7.00 - 10.00 p.m. quite separately from his daily work as an administrator.

The caution appropriate to science is seen in his comment: "This is only hazarded as a conjecture; and I trust the reader will take care not to confound what I have related as truths, fixed on the firm basis of observation and experiment, with mere hypothetical conjectures."[3] In other words, conjecture is appropriate in science but not to be confused with truths based on experimental evidence.

CHAPTER TWENTY-THREE

Notes

(1) A. Lavoisier, tr. T. Henry *Essays Physical and Chemical* 1776 Translator's preface
(2) W. Aykroyd *Three Philosophers* Heinemann 1935 p 69
(3) D. Knight *Classical Scientific Papers* Mills and Boon 1970 p 205.

24 – Triangles, chymical processes and electrical experiments

JOHN DALTON : 1766 – 1844

The King: "Well, Dr. Dalton, how are you getting on in Manchester – all quiet I suppose?"
Dr. Dalton: "Well, I don't know – just middlin I think."[1]

Whether or not the above exchange between the famous chemist and William IV actually occurred, it is certainly intriguing that such a "just middlin" weaver's son, from the then remote village of Eaglesfield in Cumbria, should "by 1808, have established the most important theoretical concept of modern chemistry".[2]

The eighteenth-century *The Ladies' Diary* and its companion for Gentlemen seem unlikely steps on the way to such a concept, but so it turned out to be. The weaver taught John and his brother mathematics and taught them so well that at age twelve John was already teaching others and before twenty was running a boarding school in Kendal with his brother. The *Diaries* presented mathematical questions to their readers and it was in the solving and then the setting of these that John made his mark.

Question No. 875 in *The Ladies' Diary* 1788 reads; "It is required to find the diameter of a circular parachute by means

CHAPTER TWENTY-FOUR

of which a man of 200 pounds weight may descend from a balloon at a great height, with the uniform velocity of only ten feet in a second of time; the parachute being supposed to be made of such material and thickness that a circle of it 50 feet in diameter weighs only 50 pounds."[3] This is hardly the sort of thing to increase magazine circulation today when apparently one in ten of us cannot multiply six by twenty-one.[4] There seems to have been no bar on men responding, and John's published prize-winning answers and his skill in problem setting led to his being invited, at age 26, to teach maths and natural philosophy at New College, Manchester.

In his native Lake District he had already been fascinated by the local flora; by weather – especially the formation of dew and the behaviour of water vapour and other gases; by the strange, flickering Northern Lights. In many years, up until his seventieth, he delighted to return there, but never just to walk. He was always experimenting with barometers, collecting gases, making observations. His sturdy fell walking prompted others to wonder what his legs were made of.

Dalton was a Quaker, as were his father and grandfather before him, which may explain the extraordinary discipline and regularity of his life. He began keeping a meteorological journal on March 24, 1787, and continued daily until the last entry on 26 July, 1844, the evening before he died. "On Sundays he always dressed himself with the most scrupulous attention to neatness and, with rare exceptions, attended public worship twice".[5] Through the period when the fancy of most young men would lightly turn to thoughts of love, John's head was "too full of triangles, chymical processes and electrical experiments etc. to think much of marriage"[6] and it seems that space never was made for such thoughts.

The idea of the atomic nature of matter was not new in

Triangles, chymical processes and electrical experiments

Dalton's day. Democritus had argued that one could not go on chopping matter into smaller and smaller pieces ad infinitum. Newton thought "that God, in the Beginning formed Matter in solid, massy, hard, impenetrable, moveable Particles."[6]

On October 21,1803, Dalton read a paper to the Manchester Literary and Philosophical Society which included the phrase, "An enquiry into the relative weights of the ultimate particles of bodies is a subject, as far as I know, entirely new. I have lately been prosecuting this enquiry with remarkable success."[7] The paper concluded with the first-ever table of atomic weights. Dalton's genius lay in combining the maths, in which he had so early been grounded, with the chemistry that later captured his interest. This enabled him to propose a system that related the weight of an atom of one substance to the weight of an atom of another and to state simple rules by which chemicals combined. It laid a quantitative base for the whole of chemical science from then on.

A Mr Ransome records an amusing incident that helps us to see how scientists tick. "On the occasion of a minor illness my father [Dalton's doctor] prescribed a small dose of James' Powder to be taken at bedtime. On the following day my father called on him again, and finding him very much better attributed the improvement to the effect of the medicine, upon which Dalton remarked, 'I do not well see how that can be, as I kept the powder until I could have an opportunity to analyse it'."[8] Is that lack of faith in his doctor, or dedication to science?

CHAPTER TWENTY-FOUR

Notes

(1) H. Roscoe *John Dalton and the Rise of Modern Chemistry* Cassell & Co. 1895 p 183

(2) F. Greenaway *John Dalton and the Atom* Heinemann 1966 p 134

(3) T. Wilkinson, in *Memoirs of the Lit. and Phil. Soc. of Manchester* 1855 Vol. 12 Second series p 6

(4) *The Times* 17/1/97

(5) W. Henry Miss Johns quoted in *Memoirs of John Dalton* Cavendish Society 1854 p 208

(6) F. Greenaway p 28

(7) F. Greenaway p 130

(8) W. Henry p 218.

25 – Armed only with a hammer and notebook

ADAM SEDGWICK : 1785 – 1873

"If he were present the gravest unbent their brows. It was impossible to resist the infection of that boisterous laugh, that cheerful geniality."[1] The geologist, Adam Sedgwick, emerges from the biographies as a jovial bear of a man. His rock-like reputation is expressed in the massive lump of granite which stands to his memory in the high street of his native Dent, in Yorkshire. There, Adam's father was the down-to-earth and much-loved parson. He was born of generations of Dalesmen in this little town, about five miles southeast of Sedbergh. Sedgwick spent most of his adult life in Cambridge, but he always thought of Dent as the "home of his heart".

By the age of nineteen, his local education had fitted him for Cambridge and he entered Trinity College in 1804. He excelled in mathematics, a tribute to his remarkable teacher, John Dawson, and won his degree in 1808. He became a Fellow of Trinity and was eventually ordained Deacon in 1817. His life then took an extraordinary change in direction. It is difficult to see what qualified him to become Woodwardian Professor of Geology in Cambridge, but he did so, and occupied the Chair for very nearly fifty-five years.

CHAPTER TWENTY-FIVE

'Geology'. The word was new-minted in France less than a decade before Adam was born. When he died, it was a well-established and many-faceted discipline. He never wrote influential books, like those on which his geological peers, such as Lyell and Murchison, built their reputations. His scientific papers were classic explanations of the observations he made while tramping around Britain, armed only with a hammer and notebook, and an alert, enquiring mind. When he died, it was proposed that a bust should be sculpted in his memory. Selwyn pleaded for a whole statue, "For what is a geologist without the hand to wield the hammer? Without the feet to carry him over the mountains?"[2] The result can be seen in the Sedgwick Museum in Cambridge, also built in his memory.

Probably his greatest achievement was his study of the oldest fossil-bearing rocks he could find, in Wales, to which he gave the name 'Cambrian'. He worked in other areas too. He became a great friend of Wordsworth, and in Hudson's *Complete Guide to the Lakes,* Wordsworth describes the scenery, and Sedgwick the geology. With Murchison he also unravelled the difficult rock formations of the West Country, giving them the name 'Devonian'. The patient putting together of the story of the rocks from these first-hand observations in the field is a vivid illustration of how science works.

Sedgwick joined the Geological Society in the year he became Professor, and little more than a decade later, was President. Sadly he became involved in a dispute with Murchison with whom he had previously worked very happily. He refers to a geological map published in Murchison's name:
" ... in which he had brushed out of sight, under a deep Silurian colour, every trace of my previous work in N. Wales."[3] The Geological Society took Murchison's side and Sedgwick recorded that "The stigma fixed upon me by the Council of the

Armed only with a hammer and notebook

Geological Society was the greatest sorrow of my old age".[4] He didn't live to see his work largely vindicated, although he remained confident that it would be.

His Cambridge lectures, begun in 1819, proved hugely popular both with Town and Gown. "Their influence on successive generations of Cambridge students and hence on the shaping of English educated opinion on geology, is hard to overestimate."[5] His eloquence has left a lasting impression. "By turns he made us cry and roar with laughter as he willed."[6] Referring to Sedgwick's farewell address at the third meeting of the British Association, a Dr Chalmers said he had " ... never met with natural eloquence so great as that of Sedgwick's"[7]. There is a remarkable account of an impromptu lecture-cum-sermon given to "three or four thousand colliers and rabble (mixed with a sprinkling of their employers)" on the beach at Tynemouth.

The annual series of lectures continued unbroken until 1870, when he was 85. His willingness to be convinced by evidence is seen almost at the end, for in the 50th series in 1868 he said, "I have gone again over the whole evidence and ... I must freely admit that man is of a far higher antiquity than that which I have hitherto assigned to him."[8]

At a meeting in Norwich, he heard a clergyman bemoaning "the influence of science in weakening belief in revelation." When the speaker finished, "Sedgwick suddenly rose, took a Bible from the table and, holding it up, exclaimed in his most vehement manner, 'Who is the greatest unbeliever? Is it not the man who, professing to hold that this book contains the word of God, is afraid to look into the other volume [the book of God's works], lest it should contradict it?'"[9] He urged that the way to resolve religious difficulties in the conclusions of geology was: " ... not by shutting our eyes to the facts or denying the

CHAPTER TWENTY-FIVE

evidence of our senses; but by patient investigation carried on in the sincere love of truth and by learning to reject every consequence not warranted by direct physical evidence."[10] On this basis he felt he was being entirely consistent in rejecting Darwin's hypothesis for, in his view, Darwin had "not added one single fact that brings it forward."[11] Others were persuaded by Darwin's evidence, but Sedgwick never was.

Indeed, he must have felt Darwin's book was mischievous, as well as unscientific, for Darwin wrote to him, "I do not think my book will be mischievous for there are so many workers that, if I be wrong I shall soon be annihilated; and surely you will agree that truth can be known only by rising victorious from every attack."[12] Even in science, one man's fact is another's fiction. This makes for important checks and balances within science and it may be a long time before consensus emerges.

In 1834 Sedgwick wrote a *Discourse on the Studies of the University of Cambridge*. It may have been the fifth edition, published in 1850, (with a phenomenal 442 page Preface!) which led to his involvement with a Royal Commission of Inquiry into the state of the major universities in 1851. In this way he met and became friendly with Prince Albert and Queen Victoria, a friendship that continued with the Queen after Albert's death. He was a guest at Osborne House on the Isle of Wight.

Also in 1834, Sedgwick became a Canon of Norwich Cathedral, which necessitated his spending part of each year in that city. His faith was a deeply personal one. Perhaps few, even in those days, would have ended a preface to a catalogue of fossils with: "And may our Maker grant that His holy light may guide the steps and warm the hearts of all who read this Preface."[13] It shows a thorough blending of the professional and the spiritual. He was vigorously against slavery and spoke out against his own university's tests which disadvantaged Dissenters.

Adam Sedgwick never married but he was a popular son, brother and uncle. In later years, a favourite niece tended him with great care and affection. At age 80 he was still able to geologise along the north Kent coast with friends, and he lectured for some years more. In the last year of his life he was increasingly confined to his rooms in Trinity College. The occasional visits he received were "like sunbeams shining through a fog". He longed for acts of social worship. At his last Advent he wrote, "I am going to begin our Sacred Service Year by celebrating, with God's permission, the Holy Communion in my own rooms today, with two, or perhaps three friends who will I hope come to support me in the holy service."[14] At the end, he was keen to hear the Psalms, 130 and 51, and John 17 with their mixture of penitence and confidence in God. He died on January 27, 1873 and was buried in Trinity College Chapel.

Some considered him "narrow". He was strongly against "Popery" but he didn't allow this to sour personal relationships and so his concern for greater unity seems genuine. "Oh that Christians would love one another, and think of the great points of common faith on which they agree, and not quarrel about the little points on which they differ."[15]

CHAPTER TWENTY-FIVE

Notes

(1) J. Clark and T. Hughes *The Life and Letters of the Rev. Adam Sedgwick* Cambridge University Press 1890 Vol. 1 Preface
(2) J. Clark and T. Hughes Vol. 2 p 483
(3) J. Salter *A Catalogue of the Collection of Cambrian and Silurian Fossils* Cambridge University Press 1873 Preface by A. Sedgwick p xxvi
(4) J. Salter p xxx
(5) ed. C. Gillispie *Dictionary of Scientific Biography* Charles Scribner's Sons NY 1981 Vol. 12 p 275 (Article by M. Rudwick)
(6) C. Speakman *Adam Sedgwick: Geologist and Dalesman* Broadoak 1982 p 89
(7) C. Speakman p 92
(8) C. Speakman p 125 J. Clark & T. Hughes Vol. 2 p 440
(9) J. Clark and T. Hughes Vol. 2 p 581
(10) C. Speakman p 104
(11) J. Clark and T. Hughes Vol. 2 p 411 (Quoted from a letter to Dr. Livingstone 16/3/1865)
(12) J. Clark and T. Hughes Vol. 2 p 359
(13) J. Salter p xxxiii
(14) J. Clark and T. Hughes Vol. 2 p 470
(15) J. Clark and T. Hughes Vol. 2 p 587.

26 – A good experiment would make him almost dance with delight

MICHAEL FARADAY : 1791 – 1867

Michael was the son of James, a Yorkshire blacksmith, and his wife, Margaret. Both were members of a small Christian sect called Sandemanians. When James moved with his young family from Outhgill to Newington Butts, then in Surrey, they enjoyed the support of another group of Sandemanians. Here Michael, the third child, was born on September 22, 1791. The father was a sick man and it seems that, for a time, the family knew extreme poverty. James died in 1810, before his son's life in science had even begun.

In 1804, after a rudimentary education, Michael left school and began delivering the newspapers that George Ribeau, a bookbinder and bookseller, hired out to his customers. This kindly gentleman took Michael on as an apprentice on October 7, 1805. His library, and the books that were bound on the premises at 2 Blandford Street, soon had the young trainee hooked on reading. Among the books he read were some which inspired a lifelong passion for science. This was helped along by his attendance at Mr Tatum's lectures, paid for by his older brother, Robert. Another influential book was by the hymn-writer, Isaac Watts. George Ribeau recorded that "Dr Watts *Improvement of the Mind,* was then read and frequent took in his Pocket." [1]

CHAPTER TWENTY-SIX

As a result, in 1809, Michael began compiling a commonplace book, one of his aims being, "to corroborate or invalidate those theories which are continually starting into the world of science."[2] Clearly, four years of spare-time reading in Mr Ribeau's library had already transformed the errand boy into a fluent writer itching to leave bookbinding for a life of science. It had also set in progress Faraday's own programme of self-improvement.

It is not difficult to imagine Michael's joy when he was given tickets for lectures by the famous chemist, Sir Humphry Davy, at the Royal Institution. Faraday took notes at these lectures, wrote them up in full, bound them carefully and presented them to Sir Humphry.

A little later Davy was temporarily blinded by one of the rather frequent explosions in his laboratory and, remembering the young Faraday, he called for him to write up his experimental notes. Faraday then applied to Sir Humphry for a position at the Royal Institution, but none was available. However, shortly after that, Sir Humphry's assistant was dismissed for brawling and Faraday was appointed in his place. These are the bones of a remarkable story which, by early March, 1813, saw Faraday established at the Royal Institution for what became his working life. But what are bones without flesh? Perhaps I can tempt you onto your own voyage of discovery through the lives of scientists by embarking upon one of the biographies.

That same year, in October, Faraday left England with Davy for a Continental tour, which lasted until April 1815. It was a revelation to one who had never been more than a few miles from London. No wonder he titled his travel diary *Curiosity Perfectly Satisfyed*. It is a delightful read. They had only got to Plymouth when Michael wrote, "Travelling I take it is fatiguing work but perhaps a little practice will enable one to bear it better."[3] Davy not only talked science, he did it as they

A good experiment would almost make him dance with delight

travelled, and Faraday met scientists on this journey with whom he corresponded for life. He also remained forever grateful to Davy for all that he learned under him on this tour and at the Royal Institution, although later the cordial relationship between them cooled.

Truth and fact were paramount for Faraday. He wrote to a friend concerning " ... the importance I early attached to facts. In reading Mrs Marcet's book on chemistry, I took care to prove every assertion by the little experiments which I made as far as my means permitted; and the enjoyment which I found in thus verifying the exactitude of the facts contributed essentially to give me a taste for chemical knowledge."[4] Ribeau's kindness had extended to allowing Faraday to set up a small laboratory at the back of the shop, where the mantelpiece had become his first laboratory bench. Faraday never forgot his indebtedness to George Ribeau. From Rome, on January 5, 1815, he wrote, "During the whole of the short eight years I was with you, Sir, and during the year or two that passed afterwards before I left England, I continually enjoyed your goodness and the effects of it."[5]

The move from mantelpiece to Royal Institution laboratory made possible the wonderful series of experiments, in chemistry and the relationship between electricity and magnetism, on which Faraday's fame rests. "According to Albert Einstein he was responsible, along with Clerk Maxwell, for the greatest change in the theoretical basis of physics since Newton."[6] Theory and experiment went hand in hand though, and the practical outcome has also been vast. "Every time you switch on a television or radio ... or travel by car, bus, train or aeroplane ... use a vacuum cleaner, washing machine or food mixer, play a cassette, video or record, think of Faraday."[7] Not that he invented any of these of course, but his fundamental discoveries made them all possible. In 1827 he began the Christmas Lectures for children. How delighted he would have

CHAPTER TWENTY-SIX

been to see them now shown on television.

In 1816 Faraday wrote the first of some four hundred scientific papers. His *Diary* is not what you might expect but, in over 16,000 numbered paragraphs, contains the exquisite record of his experiments compiled over a period of forty years. Just to handle them is to marvel at the level of dedication he gave to his self-appointed task. In Faraday's view of science, experiment was everything; and he could be patient.

On November 24, 1848 he placed a fragment of diamond in a flask of acid which was then sealed. On 24 August, 1858 he writes: "Resumed the experiments begun ten years ago ... "[8], and reports no change. It reminds us that negative results in science can be just as important as positive ones. Almost every entry in the *Diary* is a record of facts, or the scientific ideas based on them, but he once allowed himself to write, "Astonishing how great the precautions that are needed in these delicate experiments. Patience. Patience."[9] His successor at the Royal Institution, John Tyndall, recorded how a good experiment would make Faraday almost dance with delight. However, the most the *Diary* could expect was a "BUT" in capitals, thrice underlined, as he homed in on some exciting conclusion. His entry for September 18, 1845: "An excellent day's work"[10] may well, I think, be unique in the forty-year record.

Faraday's love for his wife was deep and lasting. They had been married about a year when he wrote on July 21, 1822, "Oh, my dear Sarah, poets may strive to describe and artists to delineate the happiness which is felt by two hearts truly and mutually loving each other; but it is beyond their efforts, and beyond the thoughts and conceptions of anyone who has not felt it. I have felt it and do feel it, but neither I nor any other man can describe it; nor is it necessary. We are happy, and our God has blessed us with a thousand causes why we should be so."[11] The words are in stark contrast to some he had written earlier, before Sarah swept him off his feet:

> What is the pest and plague of human life?
> And what the curse that often brings a wife?
> 'tis love.[12]

His father-in-law's comment was, "Thus does love make fools of philosophers."

The marriage was childless but Faraday loved children, and entertaining them. Life was not all long days of experiment. There are stories of him riding a velocipede along the corridors of the Royal Institution, of brewing ginger wine and making lavender lozenges in the laboratory, of throwing potassium into a bowl of water to see it whizz about on the surface in a ball of fire. There was fun and music and singing. And foundational to it all there was faith. For Faraday the Bible was not just a book; it was his guide to life. Two of his Bibles survive, marked with his emphases, notes and comments. Sundays were sacrosanct. To an intending visitor he wrote on August 23, 1839, " ... on Sundays I cannot see you being otherwise &

CHAPTER TWENTY-SIX

always engaged."[13] He was referring to the Sunday meetings with the Sandemanians.

In 1862 Michael and Sarah moved from the Royal Institution into a house, provided by the Queen, on The Green near Hampton Court. His death there, on August 25, 1867, was peaceful; his burial, in Highgate Cemetery, private, and his headstone, the plainest possible.

During the 1860s failing memory and creeping senility had gradually shrivelled his public life and dried up his correspondence, but not his "sweet unselfish disposition". In 1861 he wrote to a lifelong friend, "I am, I hope, very thankful that in the withdrawal of the powers & things of this life, – the good hope is left with me, which makes the contemplation of death a comfort – not a fear. Such peace is alone in the gift of God, and as it is he who gives it why shall we be afraid? His unspeakable gift in his beloved son is the ground of no doubtful hope; – and *there* is the rest for those who like you & me are drawing near the latter end of our term here below. ... I am happy and content."[14]

Notes

(1) L. Pearce Williams *Michael Faraday* Chapman and Hall 1965 p 12
(2) L. Pearce Williams p 12
(3) M. Faraday, ed. B. Bowers & L. Symons *Curiosity Perfectly Satisfyed* Peter Peregrinus in association with The Science Museum 1991 p 2

A good experiment would almost make him dance with delight

(4) A. de la Rive   Letter from Faraday to author, October 2, 1858, quoted in Obituary Notice in *Philosophical Magazine 34* 1867   p 410

(5) ed. F. James   *The Correspondence of Michael Faraday*   Institution of Electrical Engineers   1991   Vol. 1   p 110

(6) J. Thomas   *Michael Faraday and The Royal Institution*   Hilger 1991   p 1

(7) M. Brophy   *Pioneers of Science: Michael Faraday*   Wayland 1990   p 44

(8) M. Faraday, ed T. Martin   *Faraday's Diary*   G. Bell & Sons 1936   Vol. 7   p 332

(9) M. Faraday, ed T. Martin   1934   Vol. 5   p 228

(10) M. Faraday, ed T. Martin   1933   Vol. 4   p 277

(11) ed. F. James   Vol. 1   p 282

(12) L. Pearce Williams   p 96

(13) ed. F. James   1993   Vol. 2   p 603

(14) G. Cantor   *Faraday: Sandemanian and Scientist*   Macmillan   1991   p 81.

27 – The greatest geologist in Europe?

CHARLES LYELL : 1797 – 1875

Charles Lyell was the first-born in a family of three boys and seven girls. The birth of this "loudest and most indefatigable squaller of all the brats in Angus" occurred on November 14, 1797, at Kinnordy, the family estate in Scotland. Soon afterwards, his father took a long lease on a house near Lyndhurst, so Charles grew up on the edge of the New Forest in Hampshire. One of his childhood memories was the sight of hilltop bonfires greeting the "sad-glorious news of Trafalgar". He was at school at Ringwood and then Midhurst, but a severe attack of pleurisy kept him at home for several weeks. As strength returned, he began collecting and rearing butterflies and moths; and so developed a lifelong delight in these beautiful creatures.

From 1816 he was at Exeter College, Oxford, reading for a classical degree in preparation for a career in Law. How then did he become one of the key figures in nineteenth-century geology? There is no single answer. The New Forest, especially during holidays, and his enforced absence from school, surely fired his enthusiasm for study in the open air. The family often travelled the length of Britain, visiting the Scottish estate and friends elsewhere. Travel in those days was necessarily a leisurely business and so young Charles was exposed to the

The greatest geologist in Europe?

whole range of Britain's wonderfully varied geology and scenery. His father was a keen amateur botanist with a good library and Charles almost certainly read Robert Bakewell's *Introduction to Geology* before he went to Oxford. There he heard the stimulating geology lectures of William Buckland. Later, problems with his eyes interrupted his studies and by way of relief he travelled extensively in Europe. It was the beginning of a life of travel, all over Britain and in Europe and North America, everywhere feeding his geological passion. By 1819, he was already a Fellow of both the Geological and Linnean Societies and, although he did briefly practise Law, his election to the Royal Society in 1826 suggests that the battle had been won for geology some years earlier.

There were two outstanding features of Lyell's science. The first was his knack of searching out, and reasoning from, the evidence of the rocks and fossils. In 1831, his father-in-law-to-be wrote, "While other men content themselves with turning over dictionaries and commentaries to clear up their difficulties, nothing satisfies him but to cross the sea and break the rocks of distant mountains to clear up his geological doubts."[1] Lyell himself expressed the opinion that "A cabinet geologist can account for everything much more easily than one who takes the field, and looks all the difficulties in the face."[2]

Such bald statements do little to convey the growing excitement with which he tramped around the Scottish estate, explored the Sussex Downs and Weald with his friend, Gideon Mantell, or the French Auvergne with another friend, Roderick Murchison. A visit to Sicily made a profound impact on him when he realised that layers of rock, which he had come to believe were laid down in comparatively recent times, were *underlying* the vast bulk of Mt Etna. We see the real field-geologist at work sketching for half an hour on the summit of this ancient volcano: "The wind was so high, that the guide held my hat while I drew; but though the head was cold, my feet got

CHAPTER TWENTY-SEVEN

so hot in the cinders, that I was often alarmed that my boots would be burnt."[3] He was convinced that the same processes we see going on today, have shaped the earth in the past – a point of view he presented in his hugely popular book, *Principles of Geology*, in 1830. He revised and updated it throughout his life, so that the various editions outline the development of geology over more than forty years.

The second feature was his willingness to change his mind if he felt the evidence demanded it. He was sometimes wrong, but even then his theoretical positions " ... were always carefully reasoned; and he showed an extraordinary capacity even into old age to understand the meaning of new evidence and to change his mind."[4] For example, he accepted Darwin's account of how coral reefs are formed, although it meant giving up his own. He had become very eager to meet Darwin as reports began to reach England HMS Beagle, and when the other Charles returned in 1836, they became great friends. He was therefore well aware of Darwin's thinking long before *The Origin of Species* appeared in 1859, and it caused him much heart-searching.[5] When Lyell published his book *The Antiquity of Man* in 1863, Darwin was disappointed to find that his species theory was not more fully endorsed. It was another five or six years before Lyell was persuaded. He writes of "gradually changing my opinion" and "clinging to the orthodox faith". There were some at that time who assumed that *The Origin* implied the death of God. Lyell was not one of them. He remained, in Darwin's words, a "strong theist". The Lyells made a number of visits to America where it is recorded that they " ... always went to Church, black or white, on Sunday ... "[6]

On July 7, 1832, Charles had married Mary Horner in Germany. The marriage proved childless, but very happy. Darwin tells shamefacedly of an occasion when he and Lyell "talked for half an hour unsophisticated Geology, with poor Mrs

The greatest geologist in Europe?

Lyell sitting by, a monument of patience." However, "poor Mrs Lyell" had enjoyed a geological honeymoon and she almost always travelled with Charles on long journeys through Europe and America, and often joined him on geological outings and visits to other scientists. Early in 1839, the Lyells, with the famous botanist, Robert Brown, and others, were guests at a dinner party given by the newly-wed Darwins. Emma Darwin's comments offer a little light relief from the serious business of science. " ... in my opinion, Mr Lyell is enough to flatten a party, as he never speaks above his breath, so that everyone keeps lowering their tone to his. Mr Brown, whom Humboldt calls 'the glory of Great Britain' looks so shy, as if he longed to shrink into himself and disappear entirely; however, notwithstanding these two deadweights, viz the greatest botanist and the greatest geologist in Europe we did very well and had no pauses. Mrs Henslow has a good, loud, sharp voice which was a great comfort, and Mrs Lyell has a very constant supply of talk."[7] In old age, Charles was devastated by Mary's unexpected death in April 1873.

"The greatest geologist in Europe"? Lyell, by his books, contributed hugely to the public understanding and enjoyment of the relatively new science of geology. He also reported many original observations, sometimes with other geologists, in papers read to the Geological Society, the Royal Society, or the British Association for Science. He was knighted in 1848 and made a Baron in 1864. After he died it was said that "The great ends he accomplished were due to methodical industry, to delight in making a discovery, to love of truth, and to untiring perseverance."[8]

So far as I am aware, there is no evidence that Lyell ever thought of Christianity as a personal faith in a risen Christ, but in the twelfth edition of *The Principles,* published in the year of his death, you can read " ... in whatever direction we pursue

CHAPTER TWENTY-SEVEN

our researches, whether in time or space, we discover everywhere the clear proofs of a Creative Intelligence, and of his foresight, wisdom and power."[9]

Notes

(1) E. Bailey *Charles Lyell* Nelson 1962 p 101
(2) L. Wilson *Charles Lyell: The years to 1841* Yale University Press 1972 p 122
(3) ed. K. Lyell *Life of Sir Charles Lyell* Murray 1881 Vol. 1 p 218
(4) ed. C. Gillispie *Dictionary of Scientific Biography* Vol. 8 p 575
(5) ed. L. Wilson *Sir Charles Lyell's Scientific Journals on the Species Question* Yale University Press 1970
(6) E. Bailey p 164
(7) L. Wilson (1972) p 459
(8) ed. K. Lyell p 476, quoting D. Milne-Holme at the Edinburgh Geological Society, March 4, 1875
(9) C. Lyell *Principles of Geology* Murray 1875 12th Edition Vol. 2 p 620.

28 – What a fellow for asking questions

CHARLES DARWIN : 1809 – 1882

Down House, in Kent, is the place to go to get a glimpse of Darwin's domestic life and of his prodigious scientific output. He moved there in September 1842, with his wife, Emma, and two young children, Willie and Annie. His dream had been to escape from London and now this country retreat was to be his home for the rest of his life. But there was plenty of life before Down.

Charles was born to Robert and Susannah (née Wedgwood) on February 12, 1809, their fifth child. The father was a prosperous physician in Shrewsbury and had built himself a substantial house there, called *The Mount*. Here, Charles grew up, but in the summer of 1817, shortly after he began school, his mother died.

CHAPTER TWENTY-EIGHT

By the time he followed his brother to Shrewsbury School, the younger Darwin was already an avid collector of shells and stones.

At age sixteen Charles had little choice but to follow the Doctor's wish for him to study medicine at Edinburgh University. The wish proved abortive. Charles became engrossed in geological field studies, museum collections, marine invertebrates, freethinking and Robert Grant's version of evolution; but not medicine. He left without a degree.

Robert, finding his son no more enamoured of a career as a soldier or lawyer than as a doctor, pressed him to prepare for ordination in the Anglican Church. Charles, perhaps with thoughts of becoming another Gilbert White[1], yielded to the pressure and arrived at Christ's College, Cambridge, early in 1828, headed for the Church. This time he achieved his degree but it did not come easily. Other things happened along the way, which proved more significant, such as collecting beetles with his cousin, William Darwin Fox, or botanising with Professor John Henslow. He had realised too, that he was not "inwardly moved by the Holy Spirit to enter the Church"[2] but it was some years later before he finally abandoned thoughts of a rural parish.

Just after gaining his degree, he made a geology field trip through Wales with Professor Adam Sedgwick. He was forever grateful for Sedgwick's teaching that summer. Back in Shrewsbury after this trip, he received a letter inviting him to be travelling companion to Captain Fitzroy, and also naturalist, aboard HMS *Beagle*. Robert Fitzroy was about to sail for a survey of the coast of South America and Charles seized the opportunity to accompany him.

It was intended to be a voyage of two to three years but it lasted very nearly five, from December 1831 until October 1836. By the end of it Charles was writing "I loathe, I abhor the sea", but the experience set the course for his life's work and I

What a fellow for asking questions

can only urge you to read for yourself the accounts of his adventures by sea and overland.[3] He suffered terribly from seasickness, but his early letters home are wildly enthusiastic about the luxuriant novelty of the plants, animals and geology that came under his alert gaze. In May 1832 he wrote to Cousin William, "My mind has been since leaving England in a perfect *hurricane* of delight & astonishment. ... wandering in the sublime forests, surrounded by views I enjoy a delight which none but those who have experienced it can understand ... "[4] Throughout the voyage he was collecting specimens, making reams of notes and sending off crates of rocks, fossils, skins and plants to Henslow. When he at last arrived back at Falmouth, he made what speed he could for Shrewsbury, but the two-day journey gave him time to note "that the wide world does not contain so happy a prospect as the rich cultivated land of England." When he arrived at The Mount very late on Tuesday, 4th October, "he slipped quietly into his room exhausted, without waking the family"[5] and amazed them by his appearance at breakfast next morning.

Darwin's reputation as a geologist and collector had grown throughout the voyage, so that the first few months following his return were spent in a whirl of contacts with the scientific community of the day. Charles Lyell had been particularly keen to meet him. Gradually, specialists were found to examine the collections and report on them while Darwin turned his attention to a *Journal* of the voyage. Less than a year after his return, he began a notebook on the transmutation of species (the change of one species into another). This was the seedbed that nurtured his most famous book, *The Origin of Species.*

By late 1838, Charles was engaged to his first cousin, Emma Wedgwood, and other matters were intruding on his science. "What can a man have to say, who works all morning in describing hawks and owls; and then rushes out, and walks in a bewildered manner up one street and down another,

CHAPTER TWENTY-EIGHT

looking out for the word 'To Let'."[6] They were married on January 29, 1839 and settled in Upper Gower Street in London.

In January 1842 Darwin began hesitantly to share his thoughts on species with an unsympathetic Lyell. Later that year he pencilled a 35-page note of his theory, and two years on, now in his beloved study at Down House, he produced a much fuller account. In July, he wrote instructions to Emma, to be opened if he died, asking her to publish this account because, "If, as I believe that my theory is true & if it be accepted even by one competent judge, it will be a considerable step in science."[7]

By 1846 he had produced four books inspired by his voyage. He then turned his attention to dissecting the last of the specimens collected ten years earlier, a minute barnacle taken from the sea off Chile. The dissection launched him into eight years of research on these tiny creatures and their fossils. He eventually became as sick of them as he had of the sea all those years ago, but the books he wrote about them marked him out as a seriously competent zoologist with a very thorough grasp of the problems involved in sorting out a group of species.

While he was an undergraduate, Charles had provoked Professor Henslow into exclaiming, "What a fellow that D is for asking questions." By late 1854, with the barnacles at last behind him, he began thinking again about species in a more general way. The questions continued to flow, as they did all his life. No source of information was despised which might touch on his theory; a fact that resulted in Down House receiving surely the weirdest assortment of items in the post that has ever been delivered to one door. Darwin had long since ceased to believe that each individual species had been created separately. He saw the existing forms of life like the outer twigs of a much-branched tree, linked by other twigs and branches to a common trunk way back in time. "But why

What a fellow for asking questions

should offspring depart from their parents and go a different way? How to explain this forking?" [8] He records, "I can remember the very spot in the road, whilst in my carriage, when to my joy the solution occurred to me."[9] He knew that variation occurred – no animal or plant produced sexually is identical with either its parents or siblings. He knew there was a struggle for survival – each animal and plant may produce many more offspring than actually survive. The "solution" which gave him such joy was his theory that even tiny variations might favour an individual in its struggle to survive in the particular spot where it lived. Having survived, that variation might be reinforced in the next generation, and so on, until distant offspring of the same parents might become so significantly different from both their parents and their siblings, as to be considered new species.

Early in 1856, Lyell, in spite of his own reservations, pressed Darwin to publish. It was nearly twenty years since Charles had begun his 'species' notebook. In all that time, fearful as to how his ideas might be received, or even misused, he had only tried them out on a few close friends. His fears were, in part, justified and he was later to write, "I am so much accustomed to be utterly misrepresented that it hardly excites my attention."[10] Prompted by Lyell, he laboured at the book over the next two years and then, early in the summer of 1858, the postman delivered the bombshell that Lyell had feared. It was a paper from a young collector in Malaysia, Alfred Wallace, containing a theory so similar to Darwin's that he wrote to Lyell "I never saw a more striking coincidence."[11]

While Charles wondered how to cope with the situation, tragedy struck the Down household. The youngest child, named after his father, died of scarlet fever. With Charles and Emma distraught, two friends, Lyell and Joseph Hooker, took the matter in hand and arranged for the theories of both Darwin and Wallace to be read to the Linnean Society of London on

CHAPTER TWENTY-EIGHT

July 1, 1858.[12] The event passed almost unnoticed. Another eighteen months went by before *"On the Origin of Species by Means of Natural Selection"* became public property at the end of November 1859.[13]

Darwin lived for a further twenty-three years. They were years of patient observation, investigation and experiment in his greenhouse and garden, and resulted in ten more books. One of them was *The Different Forms of Flowers*. Most of us, probably, have not noticed anything unusual about primrose flowers. Wordsworth sums us up nicely: "A primrose by a river's brim / A yellow primrose was to him, / And it was nothing more."[14] To Darwin, the two types of flowers (have a look when you have the opportunity) were much more. "I do not think anything in my scientific life has given me so much satisfaction as making out the meaning of the structure of these plants."[15] His last book was a popular monograph on earthworms, published within a year of his death.

For most of his adult life, Darwin was dogged by debilitating illness, which, in Huxley's words, "would have converted nine men out of ten into aimless invalids".[16] When he was not completely laid low, the anodyne was work. "My chief enjoyment and sole employment throughout life has been scientific work; and the excitement from such work makes me for the time forget, or drives quite away, my daily discomfort."[17] The effects of illness were compounded by tragedy. He and Emma lost three of their ten children, but it was grief at the death of ten-year-old Annie that scarred them most deeply. Emma had the consolation of a robust faith which, as is clear from early letters, she hoped her husband would share. Charles was profoundly moved by her letters, but did not respond as Emma hoped. He left on record that "I do not believe in the Bible as a divine revelation, and therefore not in Jesus Christ as the Son of God."[18] He also wrote, "This disbelief crept over me at a very slow rate, but was at last

What a fellow for asking questions

complete."[19] Emma would probably have said that her Charles did not so much lose his faith, as never found it. Charles' grief was therefore, darkened by despair; Emma's, no less profound, was lightened by hope. There have been claims that Darwin 'recanted', or experienced a 'deathbed conversion', but the file on this subject held at Down House does not support them.

Charles died in his beloved Emma's arms on April 19, 1882. The burial was in Westminster Abbey, which upset the locals, who felt Darwin's choice would have been the parish churchyard.

Notes

(1) Gilbert White, 1720 – 1793. Author of *The Natural History of Selborne* in which he "takes his observations from the subject itself, and not from the writings of others." Everymans Library, No. 48   J.M. Dent   1949   p 113

(2) A. Desmond and J. Moore   *Darwin*   Penguin   1992   p 66

(3) ed. P. Barrett and R. Freeman   *The Works of Charles Darwin* Pickering. Vol. 1   1986   ed. N. Barlow   *Diary of the Voyage of H.M.S. Beagle*   See also Vols. 2 & 3 in the series, 1986,   *Journal of Researches*.

(4) ed. F. Burkhardt and S. Smith   *The Correspondence of Charles Darwin*   CUP   Vol 1   1985   p 232

(5) A. Desmond and J. Moore   p 195

(6) A. Desmond and J. Moore   p 274

(7) ed. F. Burkhardt and S. Smith   Vol 3   1987   p 43

(8) A. Desmond and J. Moore   p 419

CHAPTER TWENTY-EIGHT

(9) ed. P. Barrett and R. Freeman Vol. 29 1989 ed. N. Barlow *Autobiography of Charles Darwin* Collins 1958 p 145
(10) ed. F. Darwin *More Letters of Charles Darwin* Murray 1903 Vol. 1 p 25
(11) ed. F. Burkhardt and S. Smith Vol. 7 1991 p 107
(12) *Journal of the Proceedings of the Linnean Society, Zoology* 1858 p 45
(13) For the reactions to publication, see the Biographies and their book lists. E.g. Desmond & Moore as above; Jane Browne *Charles Darwin – Voyaging* Pimlico 1995; and M. White and J. Gribbin *Darwin – A Life in Science* Simon and Schuster 1995
(14) Wordsworth's *Prologue*
(15) ed. P. Barrett and R. Freeman *The Works of Charles Darwin* Pickering. Vol. 29 1989 ed. N. Barlow *The Autobiography of Charles Darwin* p 151
(16) *Nature* Vol. 25 April 27, 1882 p 597 Obituary notice
(17) ed. P. Barrett and R. Freeman Vol. 29 p 140
(18) A. Desmond and J. Moore p 634
(19) ed. P. Barrett and R. Freeman Vol. 29 p 119.

29 – The book of God's works

To ignore "The Book of God's Word", the Bible, is to court incalculable loss. Neither, however, should we ignore "The Book of God's Works", the world of nature. It is the cradle of science and of much art and a source of endless delight.

In his book *River Out Of Eden* Richard Dawkins quotes part of a letter "from an American minister who had been an atheist but was converted by reading an article in *National Geographic*".[1] The article concerned the pollination mechanisms of orchids and prompted in the reader an unexpected awareness of a creator God and a search to find out more about him. This led eventually to ordination into the Christian ministry.

Such a direct link between "The Book of God's Works" and conversion is perhaps unusual these days,[2] although Paul writing to the church in Rome, probably around AD 57, took it almost for granted. "For since the creation of the world God's invisible qualities – his eternal power and divine nature – have been clearly seen, being understood from what has been made, so that men are without excuse."[3] Certainly the great majority of scientists up until the twentieth century referred quite happily in their works to "the Creator".

Dawkins goes on to suggest that the atheist made some wrong assumptions as to how the pollinating mechanism came about and that the grounds to which he attributed his conversion were not rational. Even if this is so, the atheist *was* converted, and the process began through a sense of wonder at a marvel of nature. Paul might have said of him, " ... you have

CHAPTER TWENTY-NINE

taken off your old self with its practices and have put on the new self which is being renewed in knowledge in the image of its Creator."[4] There was a direct link between his experience of nature, even though this was at second hand in a written article, and his experience of the Creator. Conversion is often much more akin to the sudden insight of some scientific discoveries than to the step-by-step reasoning that leads to others.

The days are long past when scientists were rare birds and perhaps spent many years putting their findings into a book that then made its way around the world. There was space for anecdote, for history, for wonder, for acknowledging "The Creator". The book was probably in Latin and would be read by scholars in every civilised country, including the church leaders of the day. Now, scientists are numbered by the thousand. Their findings appear in specialist journals buried in the bowels of university buildings far from the eyes of the general public. The content may well be incomprehensible to specialists in other fields, let alone to church leaders.

A research paper in ornithology is not now likely to begin like this: "To the many good things with which the Almighty Creator has endowed man on this earth, for his use as well as for his joy, belongs the gift of birds. These have greatly appealed to my sense of beauty, and their appearance is so lovely that to my way of thinking there is nothing which contributes in a greater degree to the pleasures of the human race. What can compete with the humming bird's dazzling beauty? What can surpass in splendour the peacock's colourful tail? These are, however, not nature's only works of art deserving our attention: every kind, if we regard it with due care, presents a rich subject for admiration."[5] So wrote Linnaeus. Most modern scientists probably feel something of this wonder. Many will, in day-to-day life, happily acknowledge "the Creator". But the preferred style for today's publications

The book of God's works

is clear, concise and clinical, and probably cold. The number of readers able to find a source of wonder in any of them is inevitably very small.

By contrast, "The Book of God's Works" is open to all who will trouble to read. Most people must surely have been moved at some time or another by an unusually clear night sky or dramatic sunset or perhaps the vastness of some feature of the universe explained on television. But it can be at the other end of the scale. I have carried in my mind for years this line of poetry: "One grass blade in its veins wisdom's whole flood contains." It is poetic exaggeration of course, but for me it has always summed up the feeling that when we look at any aspect of nature there is more to it than meets the eye.

We see a blade of grass glinting in the sunlight or stirred by the wind. Put it under a microscope and we find it is made up of cells. Some of these form its outer skin and guard the millions of tiny pores that allow gases to pass in and out of the leaf. Some combine to form the spongy green tissues beneath the skin where sugar is made, the sugar upon which almost all life depends. Other cells line up to form the parcel of parallel veins running the length of the blade. Whatever its particular job each cell contains a nucleus and other parts working together to achieve the buzz we call life. The nucleus shelters pairs of minute sausage-like objects called chromosomes. The ryegrass in your lawn, for example, has fourteen pairs in each cell. The chromosomes are long strands of DNA compressed in a marvel of packaging, and the DNA is a beautifully constructed double helix, or spiral.

DNA is built up of just five different types of atom – carbon, hydrogen, oxygen, nitrogen and phosphorus, and it is this simplicity at the heart of the complexity of life that so captures the imagination. Many seem content to stop here but the poet sees a further horizon. This is not surprising. It is why poets exist. There can be no wisdom without a mind. Seeing

CHAPTER TWENTY-NINE

wisdom's whole flood in a blade of grass is to catch a glimpse of the mind of God in matter. It remains only a glimpse if it does no more than exercise our own minds or stir our imagination. However, such glimpses can open the door to faith if they set us thinking further on the mind of God. To know what is in someone's mind there must be communication, usually in words. The apostle John refers to Jesus as "the Word made flesh".[6] It is in Jesus that we encounter the mind of God most completely; and not his mind only. In Jesus, we see the heart of God too.

Notes

(1) R. Dawkins *River Out Of Eden* Weidenfeld & Nicolson 1995 p 59

(2) I would be interested to learn of other examples.

(3) *Romans* 1: 20

(4) *Colossians* 3: 9,10

(5) K. Hagberg, tr. A. Blair *Carl Linnaeus* Cape 1952 p 150

(6) *John* 1: 14.

30 – Chipping away at the confusion surrounding heat, work and energy

JAMES JOULE : 1818 – 1889

If you look at the Nutrition Information on the side of a cereal packet you will see against "Energy" a number such as 1500 kJ for every 100g of cereal you eat. I am surprised we are given this information. Surely nobody bases their choice of cereal on it: "Mummy, can we have the 1555 kJ one today?" It seems equally unlikely that breakfast is the best time to be trying to recall basic physics to work out how much of this particular cereal you need to eat to get your particular body-weight up a particular flight of stairs. Any voice suggesting the possibility of such calculations is likely to be drowned by the crunching of cornflakes. The matter is mentioned here merely because kJ stands for kiloJoule, or one thousand Joules, the units of energy named after the subject of this chapter.

James, the fourth of seven children, was born on Christmas Eve, 1818, into a well-off brewing family living at Salford, near Manchester. He had no formal school education, apparently due to ill-health, but he records that his father " ... obtained the consent of Dr. Dalton to give us [James and his older brother] 3 lessons per week in his rooms at the Literary and Philosophical Society of Manchester. Dalton possessed a rare power of

CHAPTER THIRTY

engaging the affection of his pupils for scientific truth; and it was from this instruction that I first formed a desire to increase my knowledge by original researches."[1] The instruction began in 1833, when James was fifteen, and lasted for four years. It was probably Dalton also, who suggested visits to the Lake District where, as his health improved, James enjoyed rowing and climbing, and collecting weather observations for his teacher.

At nineteen, James turned one of the rooms in his father's house into a laboratory. (His mother had died in 1834.) In the same year he produced his first scientific paper, a description of an electromagnetic engine. He hoped that such an engine would replace the steam engines of his day, but in this he was to be disappointed. In his next paper, Joule made the first recorded absolute measurement of work. We are all familiar with the words work, energy and heat, but for the physicist they have precise meanings. Work, in the scientific sense, is done when a force makes something move. If you lift something a certain distance off a table then work is done. If you lift something twice as heavy the same distance, twice as much work is done.

For ten years Joule persevered with his experiments, chipping away at the confusion surrounding heat, work and energy, but without any public recognition. However, he had confidence in his results and was no doubt buoyed up by his stated belief that "The study of nature and her laws ... is essentially a holy undertaking."[2] Early in this period he investigated the heat produced by electricity, but a paper he sent to the Royal Society was rejected and only a brief extract was printed. He was asked what he felt about the rejection. "I was not surprised, I could imagine those gentlemen in London sitting around a table and saying to each other: 'What good can come out of a town where they dine in the middle of the day?'"[3]

Chipping away at the confusion surrounding heat, work and energy

He also tackled the relationship between heat and work. For example, fire can be made by friction; a bit used for drilling may become too hot to hold; an emergency stop leaves the smell of burning rubber in the air. It was the genius of Joule to devise experiments that revealed that the heat produced is directly equivalent to the work involved, a quantity referred to as the mechanical equivalent of heat. A German physician, Mayer, had made this connection five years earlier but, as with Joule, his work had not become widely known. Independent work on laws of thermodynamics and conservation of energy was also being done by French and German physicists such as Carnot, Clausius and Helmholtz.

The years of obscurity ended in 1847 when he began a lifelong friendship with William Thomson whom he met at a British Association for Science meeting in Oxford. Joule describes this friendship as "most valuable to me in every respect, for I then found one whose whole soul was occupied with the love of truth and whose unprejudiced mind immediately entered into views which at that time had taken no hold whatsoever on the scientific world."[4] Some of those views had been given expression earlier that year when Joule gave a lecture in the reading room of St. Anne's Church, Manchester entitled, "On Matter, Living Force and Heat". It included a clear understanding of the concept of conservation of energy, but only the Manchester *Courier* saw fit to publish it, in two instalments, May 5 and 12, 1847. Big ideas do not always arrive with a fanfare.

1850 was a good year. With recognition of his work growing, Joule was made a Fellow of the Royal Society; and he and his wife, Amelia, welcomed the arrival of their first child. A daughter followed in 1852. Even such events could be turned to scientific advantage. He wrote to Thomson, "If, as I hope, you will make it convenient to be at the christening and stand godfather, we might at the same time settle the question of heat and cold from air rushing through an orifice." Their joint

CHAPTER THIRTY

work led to principles on which refrigeration was later based.

After these happy years, tragedy struck in 1854. Following the birth and death of their third child in June, Amelia became ill and, after three months of hopeful rallies and agonising relapses, she too died. He wrote, "I have lost this my dearest earthly friend ... How the loss of such a parent can be replaced to my dear children I cannot tell. I must trust in the Almighty to care for them and to direct me in their upbringing. ... It is a great satisfaction to reflect that her death had no terrors reposing as she did on the merits of her Saviour as her title to her heavenly inheritance."[5] He returned, with the children, to his father's house and never remarried.

Other house-moves followed but the one to Old Trafford resulted in problems. "My neighbours here, one of whom is an alderman of Manchester, have succeeded in temporarily stopping my researches, which they consider to be a nuisance ... "[6] One would like to know what was going on, but the neighbours may have had a point for, at one of his previous homes, the apparatus was too large for the laboratory and had to be assembled out of doors. After Joule's death there was found in the cellar of his last home, not only apparatus but also the old exercise books, written from both ends, in which he had recorded all the experiments and results. With them were loose papers, perhaps the draft of an address, in which he wrote, "After the knowledge of and obedience to the will of God, the next aim must be to know something of His attributes of wisdom, power and goodness as evidenced by His handiwork."[7]

Joule died, full of honours, on October 11, 1889.

Notes

(1) *Memoirs and Proceedings of Manchester Literary and Philosophical Society* Vol. lxxv 1930 – 31 No 8 p 110
(2) J. Crowther *British Scientists of the Nineteenth Century* Kegan Paul 1935 p 138
(3) J. Crowther p 166
(4) *Memoirs* p 113
(5) D. Cardwell *James Joule: A Biography* Manchester University Press 1989 p 159
(6) *Memoirs* p 113
(7) J. Crowther p 138.

31 – Genetics in the garden

(JOHANN) GREGOR MENDEL : 1822 –1884

The history of science is full of surprises; such as the birth of genetics in the garden of an Augustinian monastery.

The parish register records Johann's birth in Heinzendorf, Austria, on July 20, 1822; the son of Anton, a farmer, and his wife Rosine, a gardener's daughter. There is every indication that Johann's childhood was happily spent in a close-knit family in a supportive rural community that had been served for decades by its parish priest and schoolteacher. It seems that his father had an orchard, not simply for growing fruit, but for improving on the varieties available. It is not difficult to imagine how his parents' daily involvement in farming and gardening, the natural history lessons in school, and the countryside around him, all sowed the seed of his future investigations.

From his village school Johann continued his education in nearby towns up to twenty miles and more from home. Times were hard for the family and made worse by an accident suffered by Anton. Johann's younger sister, Theresia, surrendered at least part of her dowry to see her brother through the last year at high school. Just after his twenty-first birthday he entered the Monastery of St Thomas at Brunn (now Brno) and took the name Gregor.

Under Abbot Napp the monastery bustled with scientific and artistic enquiry and several monks were teachers in the

Genetics in the garden

local community. Little has come to light concerning Mendel's personal feelings or beliefs. A surviving poem shows that he saw the hand of God in creation. In some fragments of sermon notes he wrote, " ... the germ of spiritual life, sanctifying grace, is put into the soul of man ... "[1] The use of the phrase "sanctifying grace" may suggest that he understood the undeserved favour of God in providing for him a way to wholeness through Christ. But we have little to go on. Certainly the monastery seems to have encouraged a happy blend of science and faith.

On August 6, 1847, he was ordained priest, but he found the contact with illness and death unbearable and became ill himself. In the autumn of 1849 he acted as a supply teacher of maths and classics and loved it. However, when he took the teachers' exam the next year, he failed. After a period at Vienna University he was engaged, although without a degree as a supply teacher of physics and natural history at a local high school, where he taught happily for fourteen years without ever qualifying.

" ... the pupils loved him, and often visited the monastery in droves, 'like beetles buzzing at an open window'."[2] They recalled "his blue eyes twinkling in the friendliest fashion through his gold-rimmed glasses."[3] He had found his niche.

In the world of science, however, Mendel is not remembered for his attractive teaching or his devotion to his religious duties, but for a remarkable series of experiments on the hybridisation of pea plants. On arrival at the Monastery of St Thomas he had been delighted to find the botanic garden established there by Father Thaler. For two years from 1854 Mendel carefully selected pea varieties for experiment and

CHAPTER THIRTY-ONE

decided which characteristics would best suit his purpose; features such as height of plant, colour or roundness of seed. Over the next eight years he used a plot in the monastery garden, and time sandwiched between other activities, to carry out controlled crosses, harvesting, counting, measuring, and recording.

He presented his results in two lectures to the Natural History Society of Brunn in 1865, where they caused no stir. He wrote to the Professor of Botany in Munich, who failed to appreciate the significance of the letters. The lectures were published in 1866 and circulated to other countries, where, unheeded, they lingered in libraries for thirty-four years.

In *Experiments in Plant Hybridisation* Mendel states clearly what he did, meticulously tabulates his results, and shows how individual traits are not merged in the process of reproduction but can survive intact to appear in the next or subsequent generations according to simple mathematical rules. He was before his time. "Though I have had to live through many bitter moments in my life," he wrote, "I must admit with gratitude that the beautiful and the good prevailed. My scientific work brought me much satisfaction and I am sure it will soon be recognised by the whole world."[4] But it was 1900, sixteen years after his death, before workers in three different countries independently discovered that a then unknown monk had forestalled their own investigations.

He never forgot his debt to his young sister and was able to repay her generosity by making sure her three sons had a good education. This was no doubt made easier by his election as Abbot in 1868. His new responsibility left little time for science but he continued to be much loved and respected in the community.

Genetics in the garden

Notes

(1) V. Orel *Gregor Mendel – The First Geneticist* Oxford University Press 1996 p 265
(2) J.& M. Gribbin *Mendel in 90 Minutes* Constable 1997 p 30
(3) J.& M. Gribbin p 29
(4) V. Orel p 269.

32 – I shall speak to you of nothing but crystals

LOUIS PASTEUR : 1822 – 1895

June 2, 1881, was a day of high drama at Pouilly-le-Fort. Pasteur had been experimenting with a vaccine against anthrax. On May 5, he had begun inoculating half of a flock of fifty sheep. The other twenty-five were left untreated. On May 31, all the sheep were injected with a virulent strain of anthrax and a public invitation was given to see the outcome on June 2. When the crowd assembled that day, all the vaccinated sheep were alive and well, save one that later died from other causes. However, of the untreated sheep, all but three were already dead and these three succumbed before the day was out. The startling significance of what they had witnessed was not lost on the crowd, which broke into congratulatory applause. Before Pasteur set out for Pouilly-le-fort that day a telegram had told him of the success of the experiment. His wife confessed that, "I had a little moment of emotion that made me see all the colours of the rainbow."[1]

Four years later, with great trepidation, Pasteur applied a similar procedure for the first time to humans when he treated nine-year-old Joseph Meister. The boy had been badly bitten by a rabid dog and on July 6, 1885 Pasteur arranged the first of thirteen inoculations over a period of ten days. During these days, Marie Pasteur wrote to their family, "My dear children,

I shall speak to you of nothing but crystals

this will be another bad night for your father. He cannot come to terms with the idea of applying a measure of last resort to this child. And yet he has now to go through with it. The little fellow continues to feel very well."[2] Joseph survived.

It was events such as these, together with his work on fermentation, spontaneous generation and various diseases that made Pasteur a national hero and internationally famous. His method of delaying fermentation lives on in the pasteurised milk in your fridge. However, he started his scientific life as a chemist and was appointed acting Professor of Chemistry at Strasbourg just after his twenty-sixth birthday. About this time he wrote to his friend Chappuis, "Oh that you were a professor of physics or chemistry! We would work together and within ten years we would revolutionise chemistry. There are marvels hidden behind the phenomenon of crystallisation, and its study will reveal some day the intimate structure of matter. If you come to Strasbourg, you will have to become a chemist despite yourself. I shall speak to you of nothing but crystals."[3] His investigations into crystals were his most significant contribution to pure science, but that is not why most of us remember him. It is said that later in life he regretted abandoning chemical research.

Louis had been born at Dole in the Jura, France, two days after Christmas 1822, the third child of a soldier turned tanner, and a gardener's daughter. By 1827, Arbois had become the family home to which he loved to return. On his first parting

CHAPTER THIRTY-TWO

from it in October 1838 to attend school in Paris he was so overcome by homesickness he confessed to his school friend, Jules Vercel, "If I could only get a whiff of the tannery yard ... I feel I should be cured."[4] Anyone who knows what a tannery smells like will recognise this for genuine homesickness. He did eventually return to Paris for part of his education, and his doctorate in 1847 was followed shortly after by his appointment to Strasbourg. There, in May 1849, he married Marie Laurent, daughter of the Rector of Strasbourg Academy. It proved a wonderful partnership.

In 1865 Louis' father died, followed shortly after by his youngest daughter aged two and, within a year, by an older daughter aged thirteen. He and his wife had already borne the death of their eldest daughter when she was nine. Such tragedies drove him back to the old family home to recuperate but did not break his indomitable spirit. In 1868 Pasteur himself suffered severe paralysis but refused to be daunted. Twenty years later, on their wedding anniversary, his wife wrote to the surviving son and daughter, "Your father is absorbed in his thoughts, talks little, sleeps little, rises at dawn and in one word continues the life I began with him this day thirty-five years ago."[5] His life revolved around his work and was driven by a will to succeed.

In 1888, only a little more than three years after he inoculated Joseph Meister, Pasteur attended the dedication of the Institute Pasteur, founded to treat other sufferers from hydrophobia. In his speech on that occasion he contrasted the law of blood and death developing new forms of destruction, with the law of peace, work and health aiming to deliver people from the calamities that befall them. "Which of these two laws will prevail, God only knows." His own work on the germ theory of disease has brought benefit to millions. How dismayed he would have been that over a century after he died the law of blood and death still holds so many in its cruel grip

I shall speak to you of nothing but crystals

Louis Pasteur died on September 28, 1895, and was given a state funeral in Notre Dame de Paris. He was buried in the Institute which bears his name, as was Marie, fifteen years later. In 1940, Paris was in the hands of the Nazis, and some Germans wanted to visit the tombs. They found their way barred by an elderly man who refused to unlock the crypt, determined that no invader should disturb Pasteur's rest. The man was Joseph Meister.

Notes

(1) P. Debré, tr. E. Forster   *Louis Pasteur*   Johns Hopkins University Press   1998   p 400
(2) P. Debré, tr. E. Forster   p 440
(3) R. Dubos   *Louis Pasteur*   Gollancz   1950   p 34
(4) R. Valléry-Radot   *The Life of Pasteur*   Constable   !920   p 12
(5) R. Dubos   p 39.

33 – Created for science

WILLIAM THOMSON – LORD KELVIN : 1824 - 1907

"Late **again**, Sir William," squawked the parrot as its owner made yet another delayed arrival at the meal table; a remark which led to the butler being suspected of giving the parrot secret lessons. William Thomson was a scientist through and through and very likely it was work in the laboratory that kept him from dinner. J. Munro wrote, "See him engrossed in the subject of his discourse and utterly forgetful of himself, or wild with rapture over the result of an experiment, and you will say this man was created for science."[1] Thomson himself wrote, "Dullness does not exist in science."[2] It was said that he never indulged in what anyone else would have called a holiday, but there is no sense of narrowness in the records of his life handed down to us. He was a professor at the University of Glasgow for 53 years, as much loved as respected and, although his two marriages were childless, his niece paints a wonderfully warm portrait of an adored uncle.

At the heart of his scientific life were the famous "green books". His grandniece, later Mrs Ramsay MacDonald, describes them as " ... really notebooks made specially for Uncle William, which he uses up at the rate of 5 or 6 a year and which are his inseparable companions. They generally go upstairs, downstairs, out of doors and indoors, wherever he goes, and he writes in his "green book" under any circumstances. Looking through them is quite amusing; one entry will be in the train, another in the garden, a third in bed before he gets up; and so they go on at all hours of the day or night. He always puts the place and exact minute of beginning an entry."[3]

Among the calculations in the "green book" was one that led to a long-standing confrontation with T. H. Huxley, Darwin's "bulldog". It concerned the age of the earth. Thomson realised that, unless some new source of heat was discovered, the cooling of the earth must give a clue to its age and he went on to work out, within limits, what this must be. The result did not please those, including Darwin and Huxley, for whom the theory of evolution appeared to require a vastly greater expanse of time. Huxley referred to Thomson as a " ... passer-by, who fancies our house is not so well built as it might be." Thomson's reply gives the measure of the man: "For myself, I am anxious to be regarded by geologists, not as a mere passer-by, but as one constantly interested in their grand subject, and anxious in any way, however slight, to assist them in their search for truth."[4] Just a few years before Thomson died his calculations became obsolete because a new source of heat *was* discovered, in radioactivity.

William was born in Ireland in June 1824, one of six children who survived infancy. His mother died when he was six and the motherless family moved to Glasgow in 1832 when his father James was appointed professor of maths at the university. In spite of a precocious talent, " ... There never was a more boyish boy or one more full of fun."[5] He was taught at home, entered St Peter's College, Cambridge, and published several mathematical papers before he graduated. By the age of 22 he was Professor of Natural Philosophy at the same university as his father. "His original methods of demonstrating his theories were always a source of unfailing delight to

CHAPTER THIRTY-THREE

his students. ... irrepressible applause was excited when he brought out his old Jacob Volunteer rifle and fired it at the heavy bob of a ballistic pendulum to show the displacement produced by the impact of the bullet."[6] One wonders where this demonstration took place! When he retired 53 years later he registered as a research student, scientifically active to the end of his life.

At an oration celebrating the centenary of his birth it was said that "His name would be world famous if he had written on only one of the following subjects: thermodynamics, electricity, magnetism, elasticity, telegraphy, heat, hydrodynamics, electrical engineering, maths, dynamics or navigation ... His works are a gift to universal humanity."[7] But his fame did not rest simply on his scientific writing. His numerous inventions included a compass adopted throughout the navy, and he was knighted for his part in laying, after several attempts, the first Atlantic cable. His involvement in this and other cable-laying exploits make a gripping adventure in themselves and led to the tragedy of his nephew's death at sea and the romance of his second marriage. The death of his first wife in 1870 had left him desolate.

On New Year's Day, 1892, during his presidency of the Royal Society, he was made a peer of the realm. This caused much merriment in the family circle. "There was great fun and excitement over what the name was to be. We were full of brilliant ideas: Lord Compass, Lord Cable, Lord Netherhall [the name of his home in Largs]. Uncle William suggested Lord Tom-Noddie would be rather nice."[8] Kelvin, the name of the little river running through the university grounds, eventually won the day.

When Lord Kelvin died at Netherhall on 17 December 1907, it was written of him, "He was a sincere Christian as taught by Christ, not by the churches ... "[9] It is sad, if unsurprising in some cases, when such a distinction is made. Paul reminds us that "Christ loved the church and gave himself up for her to make her holy, ... "[10] People have a right to expect that the teaching of Christ and of the churches should be one.

Notes

(1) J. Munro *Bijou Biographies: Lord Kelvin* Henry J Drane 1902 p 104
(2) W. Thomson *Popular Lectures and Addresses* McMillan & Co 1891 Vol. 2 p 491
(3) A. Gardner-King *Kelvin the Man* Hodder & Stoughton 1925 p 121
(4) W. Thomson Vol. 2 p 112/113
(5) A. Gardner-King p 7
(6) A. Gardner-King p 19
(7) *8 Pamphlets on the History of Physics* Glasgow University 1924 p 21
(8) A. Gardner-King p 105
(9) A. Gardner-King p 158
(10) *Ephesians* 5: 25,26.

34 – Among the greatest of all intellectual achievements

JAMES CLERK MAXWELL : 1831 – 1879

James was born at 14 India Street, Edinburgh, on June 13, 1831. "He seems to have been one of those children who are always into things, trying to find out how they work, and what *else* they can be used for."[1] His father encouraged his restless questioning: "What's the go of that?"; "What does it do?"; "Yes, but what is the *particular* go of it?"[2] If you have anything to do with a child like that, keep in mind what a little encouragement did for James and therefore for science.

James' mother died when he was only eight, a devastating event that may have further strengthened the extraordinary closeness between father and son. When James set off for the Edinburgh Academy, he went in clothes designed by his father – not a good idea as it turned out. They marked him as something out of the ordinary and may have contributed, with other eccentricities, to his school nickname – "Dafty". At first he didn't shine at school but then, quite suddenly, began to win prizes in mathematics and English verse. From such unlikely beginnings emerged one of the most brilliant of all scientific minds.

While only fourteen he had mathematical papers accepted by the Royal Society of Edinburgh. From the Academy he went on to Cambridge and became a Fellow of Trinity College in

Among the greatest of all intellectual achievements

1855. Aged 25, he returned north on his appointment as Professor of Natural Philosophy at Marischal College in Aberdeen. Four years later he filled a similar post at King's College, London, where the years 1860 – 1865 were his most fruitful. He then spent a period back at Glenair. This was the home built by James' father on the family estate near Castle Douglas. 1871 found him back at Cambridge as Professor of Experimental Physics. Here he established the Cavendish Laboratory and, just prior to his early death in November 1879, he published an account of the electrical researches of Henry Cavendish. This, in a nutshell, is the record of his brief, but full life. So – why is he famous? Because there was "scarcely a branch of the physical sciences in which he had not made his mark and he had completely transformed four, namely colour vision, the kinetic theory of gases, electricity, and magnetism."[3]

For most of us, the trouble with science is that it is seldom immediately accessible to us to the extent that art or music is. We may simply not understand the equations on which Maxwell's fame is built. But we can listen to others who do. "While in the kinetic theory of gases, Maxwell shared his leadership with several others, in the field of electrodynamics his genius stood alone. For to him was given, after many years of quiet investigation, a success which must be numbered among the greatest of all intellectual achievements.

By pure reasoning he succeeded in wresting secrets from nature, some of which were only tested a full generation later, as a result of ingenious and laborious experiments."[4] Or again; "The whole field of optics, which had defied attack from the side of mechanics for more than a hundred years, was at one stroke conquered by Maxwell's Electromagnetic Theory. Every optical phenomenon can now be treated as an electromagnetic problem. This must remain for all time one of the greatest triumphs of intellectual endeavour."[5] Building on

CHAPTER THIRTY-FOUR

the experimental foundations laid by Faraday, Maxwell handed on a tool which in the hands of Hertz, Lodge and others led to Marconi and the communications revolution.

The talent for English verse discovered at school led on to the writing of several poems including, at the age of 21, *A Student's Evening Hymn* which makes clear both his wonder at Creation and his confidence in God. His wife Katherine, whom he married in 1858, shared his Christian faith and devotedly nursed him through several illnesses. In September 1865, at James' insistence, she read to him daily their portion of Scripture, even though his illness was such that this was "the utmost mental effort he could bear."[6] "It is said that at 8 years old he could repeat the whole of the 119th Psalm [the longest in the Bible at 176 verses]. His knowledge of Scripture from his earliest boyhood was extraordinary, extensive and minute ... "[7]

Certainly his letters to his wife reveal a deep familiarity with the Bible. Concerning Jesus, he once wrote to her, "I can always have you in mind – why should we not have our Lord always before us in our minds, for we have His life and character and mind far more clearly described than we can know anyone here? If we had seen Him in the flesh we should not have known Him any better, perhaps not so well. Pray to Him for a constant sight of Him, for He is man that we may be able to look to Him and God so that He can create us anew in His own image."[8]

Maxwell was a man of faith "too deep to be in bondage to any set of opinions". He did not find it necessary for his science to deny creation. "Why should not the Original Creator have shared the pleasures of His work with His creatures and made the morning stars sing together?"[9]

Notes

(1) J. Hendry *James Clerk Maxwell* Hilger Ltd 1986 p 108
(2) *Essays on James Clerk Maxwell. A Commemorative Volume 1831 – 1931* J. J. Thomson C U P 1931 p 2
(3) J. Hendry p 255
(4) *Essays* Max Planck p 56
(5) *Essays* Max Planck p 57
(6) L. Campbell & W. Garnett *The Life of James Clerk Maxwell* McMillan & Co 1882 p 320
(7) Campbell & Garnett p 32
(8) Campbell & Garnett p 339
(9) ed. P. Harman *The Scientific Letters and Papers of James Clerk Maxwell* CUP 1990 Vol. 1 p 228.

35 – Electrified by his own enthusiasm

WILHELM RÖNTGEN : 1845 – 1923

News of the discovery of X-rays was greeted with enormous public excitement. Their discoverer was fêted in a way few scientists have ever been. In 1895 Röntgen was Professor at the Physical Institute in Würzburg, Germany, a well-established physicist with a wide circle of scientific contacts. At fifty years of age, he already had forty-eight papers to his credit, mostly to do with heat and crystals. During the summer of that year, he turned his attention to something that was fascinating many other scientists at the time.

The invention of efficient vacuum pumps and vacuum tubes had made it possible to study how electricity behaved in a vacuum, or in gases at low pressure. When heavy electric currents were passed between terminals sealed into the vacuum tubes, beautiful coloured light effects were produced, varying with the pressure and what gas was in the tube. Sometimes the glass itself would fluoresce brilliantly. The German physicist, Plücker, showed that some kind of radiation travelled in straight lines from the negative terminal (the cathode). The rays were called cathode rays and could be deflected by a magnet held near the tube. Röntgen was intrigued and, armed with the appropriate apparatus, set to work. He had been busy for some months when, on November 8,

Electrified by his own enthusiasm

something totally unexpected happened. For his experiments that day he had completely blacked out the tube, but noticed that a sensitive screen which happened to be lying on a bench some distance away was fluorescing brightly. Knowing that any effects of cathode rays vanished within a short distance from the tube, he realised something quite different was at work.

For the next six weeks he was reluctant to leave his laboratory, adjacent to his living quarters, and his taciturn behaviour left his bewildered wife wondering what could be going on. He later wrote, "To my wife I mentioned merely that I was doing something of which people, when they found out about it, would say, 'Röntgen has probably gone crazy'."[1] He later explained his behaviour, saying, "I made the observations many many times before I was able to accept the phenomenon myself. During these trying days I was in a state of shock."[2]

The biggest shock was to see the bones in his own fingers revealed, as he held different objects in the path of these mysterious rays. Just before Christmas, Wilhelm took his wife into the laboratory, and, to show her what he had been up to, took an X-ray photo of her hand. It was not a good idea. What he intended as a Christmas surprise, seemed, to his frightened Bertha, a gruesome spectacle. During those weeks of feverish and lonely experimenting, Röntgen had worked out pretty well all that was to be known about X-rays for over a decade. He set it all down very concisely in a paper, "On a New Kind of Rays" which he presented to the President of the Physical and

CHAPTER THIRTY-FIVE

Medical Society of Würzburg on December 28. It was immediately published and on January 1, 1896, Röntgen was able to send reprints, along with samples of X-ray photographs, to several European scientists. The response was dramatic.

The scientific world offered amazed congratulations. The medical potential for these new rays was immediately grasped and was greeted by the press with an orgy of enthusiasm and inaccuracy. Invitations to make personal appearances to receive honours arrived by every post. Five editions of his original paper were called for within three months. Any hope of continuing his experiments was washed away in the flood of excitement, and this in spite of the fact that he steadfastly refused all invitations and interviews. He made one exception to each. He interpreted as a summons, the Kaiser's invitation to the palace to demonstrate the new rays, and this took place on January 13. One lucky journalist was granted an exclusive interview. Describing Röntgen he wrote, "his long dark hair stood straight up from his forehead, as if he were permanently electrified by his own enthusiasm."[3]

The tribute which moved him most deeply was the appearance of a band-playing, torchlight parade of the student body which gathered outside the Institute one evening to honour the Professor. He thanked them from the vantage point of his first floor window. No wonder Frau Röntgen wrote to a friend, " ... the work was barely published when our domestic peace was gone completely." She was not unhappy, however, and added, "We are often almost dizzy with all the praise and honours bestowed on him."[4]

The lack of domestic peace was not the only drawback. The discovery stirred up the usual cloud of cranks and beggars that whirls around such events for a while. It was presumed by some that the penetrating power of the rays would do away with the need for vivisection. It was claimed that X-rays "would do much for the temperance cause" now people could see for themselves "the steady deterioration of their systems". Suppliers offered X-ray opera glasses; others countered with advertisements for X-ray-proof underclothing. Many Christians were startled to learn that Röntgen's rays "proved the very existence of a spiritual body in man". If that were so, surely it must be true also for the cats and dogs, fish and frogs which so quickly appeared in skeletal photographs. A young servant girl asked B. Hunter of *The Lancet* if he would " ... look through her young man unbeknown to him ... to see if he was healthy in his interiors."[5] Cartoons and poems enlivened the popular press. "I'm full of daze/Shock and amaze; /For nowadays/I hear they'll gaze/ Thro' cloak and gown and even stays/ These naughty, naughty Roentgen Rays."[6] The discoverer himself "complained that he was about to wear out a perfectly new hat from returning the greetings of people whom he did not know when taking his daily walks."[7]

More seriously, prior claims to the discovery were made, but none have been confirmed. Röntgen has been described as a person of 'absolute integrity' and he took care to acknowledge those whose work contributed to his own. This quality is also seen in his attitude to financial gain. "Representatives of American Companies were the first to attempt to buy my discovery by holding millions before my eyes." A German Electrical Company also wanted him to seek a patent and work with it, but, " ... according to the good tradition of the German university professors, I am of the opinion that their discoveries and inventions belong to humanity and they should not in any way be hampered by patents, licences, contracts, or be controlled by any one group."[8] If only the modern world would hear that. In the area of genetic discovery there surely are huge potential dangers in the failure to hold to that ideal.

So far in this brief account, as in real life, the man has been lost in his discovery. He was born in Lennep, near Cologne, on

CHAPTER THIRTY-FIVE

March 27, 1845, but spent most of his childhood in his mother's native Holland. He nearly missed a university education, but eventually studied in Zurich, where he met the love of his life, Bertha Ludwig. With no clear goal in life, Wilhelm was happy to follow his professor's suggestion to take up physics and to move with him to the university at Würzburg. There Wilhelm and Bertha were married in 1872. At first there was no family, but several years later they fostered and eventually adopted a young niece of Bertha's.

In due course, Röntgen became Professor of Physics in turn at the universities of Geissen, Würzburg, and Munich. The happiest period was their second stay at Würzburg to which they had returned in 1888. Here they attended the Protestant St Stephen Kirche rather than the large university church. They had many close friends, with some of whom they shared a great joy in nature and often spent long vacations. Wilhelm was an enthusiastic mountaineer, hunter and botanist. Husband and wife shared a particular delight in a hunting lodge they had built in beautiful scenery south of Munich. Their visits there became more frequent and longer, as age and infirmity made more distant visits difficult.

Röntgen wrote and spoke very little on religion. The couple had a Bible, given them as a wedding present, from which Wilhelm often read portions aloud to Bertha. She suffered from recurring bad health, but even so, lived to be 81, helped by Wilhelm's devoted nursing. In 1901, she had been too ill to accompany him to Stockholm when he received the first Nobel Prize, for physics. During her last illness, they were forced to move house and this probably hastened her death, which occurred after only a few days in the new home. Desolate and grieving, he was helped by friends. To one he wrote, "'God protect you' were the words my wife frequently spoke to me during her last days. You are the instruments in His hands to fulfil this prayer, and He has chosen the best that He alone

Electrified by his own enthusiasm

could give."[9]

The war had saddened their last years together but Bertha was spared the worst of the terrible privations of the post-war period. Wilhelm was invited to live in Switzerland but refused to live in comfort while the nation suffered. In 1921 he was already deploring signs of anti-Semitism. By his death on February 10, 1923, he escaped the horrors to come.

In 1959 Professor Arthur Compton wrote, "We can show that the number of lives that have been saved by X-rays since their discovery by Röntgen is as great as the number of lives that have been taken in all of the wars that have been fought since that time."[10] What a legacy for the man who resisted all attempts to name his discovery 'Röntgen Rays'.[11]

Notes

(1) W. Nitske *Wilhelm Conrad Röntgen* University of Arizona Press 1971 p 100
(2) W. Nitske p 5
(3) W. Nitske p 130
(4) W. Nitske p 139
(5) O. Glasser *Wilhelm Conrad Röntgen* Bale 1933 p 45
(6) O. Glasser p 44
(7) W. Nitske p 139
(8) W. Nitske p 174
(9) W. Nitske p 267
(10) W. Nitske p 259.
(11) In spite of this, Röntgen Rays is the name which the rays permanently acquired in Germany, where the verb 'rönchen' means 'to X-ray'.

36 – The search for truth

Pilate's famous question, "What is truth?"[1] still echoes down the centuries. For science, the search for truth is the pursuit of reality in the universe. What are things really made of? How do they tick? It is not content with appearances. It wants to know what happens.

Lord Kelvin's long life in science illustrates some interesting aspects of the search for truth. He wrote, "The search for absolute and unmistakable truth is promoted by laboratory work in a manner beyond all conception."[2] This is fine so long as we remember that the nuggets of truth gathered in this way may be melted down and reshaped by the next generation of laboratory work. An individual nugget may, or may not, keep its identity but all contribute to the mass of gold which is the body of truth accumulated by science. If he were alive today, Kelvin might hesitate to use the word "absolute" in this context. Science is cumulative: neither standing still, nor, probably, ever complete.

The experiments of James Joule, together with his own work on heat, enabled Lord Kelvin to calculate an approximate age for the earth, assuming there was no other, as yet undiscovered, source of heat. The implications of this calculated age so worried Darwin that he referred to Kelvin as "an odious spectre". The laboratory work had promoted an "unmistakable truth" which, on the knowledge available to him, Darwin could not gainsay, much as he would like to have done

The search for truth

for the sake of his theory. Then, on March 16, 1903, everything changed. Curie and Laborde announced their discovery that radium salts constantly release heat. It was the answer to Darwin's worries that he never lived to see and that Lord Kelvin never publicly accepted. It seems paradoxical that scientists, who of all people should be most open-minded, often find it so difficult, especially late in life, to accept a revolution in thinking.

While digging for information about the scientists who make up this book, I have been intrigued and delighted to turn up something of the lives behind the names. I have been intrigued to know where these people came from and, in some cases, what sparked off the scientific adventure. I have been delighted to catch something of their enthusiasm and perseverance, unearthing nature's secrets as they pursued their search for truth. Several of them are men or women of faith. Some are happy to confess to a personal belief in the risen Christ as Saviour and Lord. Their lives are evidence that the pursuit of science, even at the very highest intellectual levels is no barrier to faith. Where there is apparent conflict between the two it is better to be patient than to denounce one or the other.

In the case of science, accumulating evidence will either demand a change in the scientific explanation, or it will tend to confirm it. If the latter, then even long-held and cherished beliefs about how the world works may have to yield. When Galileo's telescope revealed mountains on the moon, many were aghast. How could it be? The moon was a heavenly body and *therefore* a perfect sphere. Other observers confirmed what Galileo had seen and, in this case agreement on the evidence was reached quite quickly. Science is forever discovering 'mountains on the moon' which pose a challenge to the way we think about the universe. If we accept the validity of science at all then we cannot pick and choose the evidence, welcoming that which supports our belief but dismissing that which does not. Certainly we may keep an open mind, remembering that

CHAPTER THIRTY-SIX

the history of science is full of surprises.

The more one explores that history and the lives of the people who made it, the deeper grows the conviction that the great sweep of science is a search for truth by men and women of integrity. Certainly there has been resistance to evidence; a clinging to familiar thinking, but that may simply make people search the harder. Certainly there have been some who cooked the books to try to further their own reputation, but, in the end, such reputations are destroyed, not enhanced. For there seems to be a certain inexorability about the scientific process by which, sooner or later, the truth will out. Science may be abused by partisan regimes for their own purposes. It may be turned by the unscrupulous to evil ends. But in itself it is neutral, serving only to reveal the secrets of the Universe and to say, with ever-increasing exactness, "This is how it is".

When it comes to faith, the evidence may not be something that can be weighed or measured, but millions have found it compelling. Nearly two thousand years ago Peter and the other apostles were brought before the supreme Jewish court. There were those who wanted the apostles killed, but the advice of Gamaliel prevailed. "Let them go! For if their purpose or activity is of human origin, it will fail. But if it is from God you will not be able to stop these men; you will only find yourselves fighting against God."[3] From that small group of Christ's followers has sprung today's Church, now worldwide and still growing.

In the discovery of DNA it was, as we shall see, the elegance of the model produced by Crick and Watson that convinced colleagues that they had got it right. The compelling evidence was in the detail but the model had about it a self-evident authority or truth. I believe we see something similar in the life of Christ. "The people were amazed at his teaching, because he taught them as one who had authority, ... ".[4] Again, the compelling evidence was in the detail of his life and teaching,

The search for truth

his death and resurrection, but there was about his person such openness and transparency, such integrity and directness, that people saw, and still do see, a self-evident authority.

Today, the evidence for the reality of faith lies in the lives of those who reflect the life of Jesus most closely. Christians <u>are</u> being transformed into his likeness[5], but if we look in the local church for such evidence we may be disappointed. This is because every church is a bunch of sinners, some already saved by grace, but all still being worked on by the Holy Spirit, and none perfect. The process of being changed is not a comfortable one and too many of us who claim to be Christian have other priorities. We prefer to maintain the authority of the Church, defend a particular doctrine, insist on our interpretation of Scripture, busy ourselves about the Lord's work; anything rather than to accept the biblical priority of love.[6] Not only this, but sometimes we push our own priorities with such arrogance and lack of grace that we betray the very name of the One we claim to serve.

As with science, Christianity too can be abused by the partisan and the bigot, or turned to evil ends, but, sooner or later, these will be seen for what they are, a betrayal of him who said, "I am the truth." As we search for truth we need to realise that it is not doctrine that gives Christianity its authority. It is a Life: a Life given as a ransom for many.[7]

CHAPTER THIRTY-SIX

Notes

(1) *John* 18: 38
(2) W. Thomson *Lectures and Addresses Vol. 2 Geology and General Physics* Macmillan 1891 p 489
(3) *Acts* 5: 38,39.
(4) *Mark* 1: 22
(5) *2 Corinthians* 3: 18
(6) *1 John* 4: 7-21
(7) *Mark* 10: 35-45.

37 – Engraved on the heart of theoretical physics

MAX PLANCK : 1858 – 1947

If science would only confine itself to apples falling from trees, more of us would probably be happy to keep it company. However, when words like "quantum" appear on the destination board even highly educated minds have hesitations about the journey. Planck's name is engraved on the heart of theoretical physics but it is given to relatively few to appreciate fully why this is so. We may or may not understand quantum theory but none of us is denied the inspiration to be gained by seeing how his spirit triumphed over tragedy.

Planck's earliest studies were in radiation. I doubt if many of us, when holding our hands to the warmth of a fire, give much thought to how the heat reaches us. Planck spent the first twenty years of his scientific career wrestling with problems in radiation which classical physics had not been able to answer. Eventually, in his own words, " ... after some weeks of the most intense work of my life clearness began to dawn upon me, and an unexpected view revealed itself in the distance."[1] He arrived at a new radiation formula and, at the turn of the century, came to the conclusion that, "radiant heat is not a continuous flow and indefinitely divisible. It must be defined as ... made up of units all of which are similar to one another."[2] Each unit he called an "elementary quantum of

CHAPTER THIRTY-SEVEN

action" and, even more remarkably, derived a value for it. It marked the birth of quantum physics. Niels Bohr wrote, "Scarcely any other discovery in the history of science has produced such extraordinary results within the short space of one generation."[3] Theoretical physics was transformed.

What of the man? Max Planck was born in Kiel, Germany. The family background was the law and the Protestant ministry. He grew up to revel in mountain climbing and music. He might even have become a professional musician, but when he sought advice about following a musical career he was told, "If you have to *ask*, you'd better study something else!"[4] The "something else" turned out to be physics. After education at the universities of Munich and Berlin he was appointed to the staff of Kiel University in 1885, and three years later to Berlin where he stayed until 1926.

From the age of 50 his life was peculiarly seared by tragedy. The mere chronicle of it staggers the mind. His first wife died in 1909 followed, during the First World War, by three of their four children; his eldest son in action and his two married daughters in childbirth. A happier interlude produced a son from his second marriage. In the Second World War, as president of the Kaiser Wilhelm Gesellschaft, he met only rebuff for his efforts to influence Hitler against the worst of his evils. He escaped with his life but the only surviving son of his first marriage was executed by the Nazis for conspiracy against Hitler. Before the war ended Planck, now in his eighties, lost his home in an air raid and with it all his possessions, including his precious library. As Germany emerged from barbarism there were plans to celebrate his ninetieth birthday but he died six months earlier on October 4, 1947.

Planck became a churchwarden in Berlin after the First World War and it is recorded that he "professed his belief in an almighty, omniscient and beneficent God".[5] It gives food for

Engraved on the heart of theoretical physics

thought that a rational scientist of Planck's calibre could suffer as he did and speak of a "beneficent God". Certainly there were aspects of traditional Christianity which he did not accept but he wrote widely and with fascinating insight on such topics as causality and free will, and on the relationship between science and religion. The following quotations give some idea of his thinking. "Religion and natural science do not exclude each other, as many contemporaries of ours would believe or fear; they mutually supplement and condition each other. The most immediate proof of the compatibility of religion and natural science, even under the most thorough critical scrutiny, is the historic fact that the very greatest natural scientists of all times – men such as Kepler, Newton, Leibnitz – were permeated by a most profound religious attitude. ... Religion and natural science are fighting a joint battle in an incessant, never relaxing crusade against scepticism and against dogmatism, against disbelief and against superstition, and the rallying cry of this crusade has always been, and will always be: 'On to God!'."[6] And again, "There can never be any real opposition between religion and science; for the one is the complement of the other. Every serious and reflective person realizes, I think, that the religious element in his nature must be recognised and cultivated if all the powers of the human soul are to act together in perfect balance and harmony."[7]

CHAPTER THIRTY-SEVEN

Notes

(1) M. Planck *The Origin and Development of the Quantum Theory* Oxford/Clarendon 1922 p 9
(2) M. Planck *Where is Science Going?* George Allen & Unwin 1933 See introduction by James Murphy p 21
(3) M. Planck *Where is Science Going?* p 19
(4) ed. C. Gillispie *Dictionary of Scientific Biography* Charles Scribner's Sons 1981 Vol. 11 p 7
(5) ed. C. Gillispie Vol. 11 p 14
(6) M. Planck, tr. F. Gaynor *Scientific Autobiography* Philosophical Library New York 1949 p 186/187
(7) M. Planck *Where is Science Going?* p 168.

38 – Science has great beauty

MARIE CURIE: 1867 – 1934

The life of Marie Curie is more than the story of a rather plain woman in a black dress spending years in a cold laboratory on the way to a double Nobel Prize.

Her life began on November 7, 1867, as Maria Sklowdowska, in a Poland with no national identity, divided between Austria, Prussia and Russia. The attempt by Russia to weed out everything Polish led to a clandestine double life in school, and all the petty restrictions of an oppressive regime. Childhood was overshadowed by the deaths of her eldest sister, and of her mother before Maria was eleven. However it was not all gloom. She enjoyed memorable times making visits to relatives in the countryside which gave her a life-long love of nature.

Some bleak years followed when she took a post as governess to enable her older sister, Bronia, to attend the University of Paris. These years included a broken romance. Matters improved when she was able to return to Warsaw for a

CHAPTER THIRTY-EIGHT

year and live with her father and, " ... to my great joy to be able, for the first time in my life, to find access to a laboratory."[1] Not everyone associates laboratories with joy, but for Maria they became the hub around which her life revolved.

In November 1891, with Bronia now married and established in Paris, it was Maria's turn to study at the University there. She immersed herself totally. "All that I saw and learned that was new, delighted me. It was like a new world opened to me, the world of science, which I was at last permitted to know at all liberty."[2] Also at the University were two brothers, Pierre and Jacques Curie. It was Pierre who fell for Marie, as she now became known. "One of Pierre's first billets-doux to Marie was a copy of his 1894 paper on 'Symmetry in physical phenomena'."[3] This may not seem the stuff of romance but they married in July 1895, and cycled off on honeymoon on the shiny new bicycles they had given each other as wedding presents. Life now became a blend of discovery and domesticity.

In 1896, Henri Becquerel found that uranium compounds gave off strange rays. Marie decided to investigate. She worked in a storage space on the ground floor of the school where Pierre now taught. Their joint efforts led to the discovery of a new element, and the first use of the word 'radioactivity'. Their paper, read on July 18, 1898 was titled, "On a new radioactive substance contained in pitchblende". They called it Polonium after Marie's homeland. At about the same time she noted in the margin of her cookbook the making of a large batch of gooseberry jelly. "I obtained 14 pots of very good jelly, not transparent, which 'took' perfectly."[4] She was about to discover that making jelly was a very simple task compared with isolating a sample of radium, which was the next element they found in pitchblende. "Sometimes I had to spend a whole day mixing a boiling mass with a heavy iron rod nearly as large

Science has great beauty

as myself ... "[5]

By July 1902 she had managed to extract one tenth of a gramme of radium from about eight tons of pitchblende! "It had taken me about four years to produce the kind of evidence which chemical science demands, that radium is truly a new element."[6] Sometimes the Curies returned to their laboratory in the evening to view their precious products. " ... their faintly luminous silhouettes all around us, and these glowing lights, which seemed to be suspended in the darkness, always thrilled and delighted us all over again."[7]

The Nobel Prizes were awarded for the first time in 1901. In 1903 the physics prize was divided; half going to Becquerel for " ... the extraordinary services he has rendered by his discovery of spontaneous radioactivity", and half to Pierre and Marie for " ... their joint researches on the radioactive phenomena discovered by Prof. Henri Becquerel."[8] The award inspired overwhelming media interest, not to mention commercial hangers-on and any amount of quackery. George Bernard Shaw complained that the world had gone radium mad. It amazed Pierre that journalists had "gone so far as to reproduce the conversations of my daughter with her maid."[9] Irène had been born in September, 1897 and another daughter, Eve, followed in December 1904.

Then, in 1906, after a happy Easter weekend in the country, death dealt the family a devastating blow. Pierre fell under a heavy wagon and was crushed. Marie's tragic journal, into which she poured out her grief, was only made available in 1990. " ... never will I have enough tears ... " It was as well that Pierre's father lived with the family. His serenity helped to prevent the children, especially Irene, foundering in the ocean of Marie's grief. But four years later, he too was dead.

Marie steeled herself to continue the work, and was offered, and accepted Pierre's professorship. It was an exciting time in science. Thomson and Rutherford in England were showing

CHAPTER THIRTY-EIGHT

that the atom was not solid and indivisible. The discovery of X-rays and radioactivity had astonished the community of physicists. Marie herself was awarded a second Nobel Prize in 1911, for chemistry this time, for " ... the discovery of the elements radium and polonium, the isolation of radium, and the study of the nature and compounds of this remarkable element."[10]   However, there was no repeat of the glory associated with her first Prize, for the heroine of 1903 was now hounded by the press over the scandal of her association with a married colleague, Paul Langevin. The result, as usual, was a lot of pain for a lot of people. It was probably only Marie's vigorous war effort in setting up dozens of permanent and mobile X-ray units near the front line that brought her back onto an even keel and reconciled her to the public.

After the war, Irène joined her mother in the laboratory as Assistant. She married a fellow Assistant, Frédéric Joliot. Their discovery of artificial radioactivity was one of the last great excitements of Marie's life. Her son-in-law later recalled "the expression of intense joy that came over her face"[11], as they showed her their discovery. In 1935, they too were awarded a Nobel Prize, but on July 6, 1934, Marie had been laid to rest in the little cemetery at Sceaux, beside her beloved Pierre.

Few lives have been more dedicated to science. She enjoyed the thrill of discovery and wonder at what she had discovered. "I believe", she wrote, "that science has great beauty. A scientist in his laboratory is not a mere technician: he is also a child confronting natural phenomena that impress him as though they were fairy tales. We mustn't let anyone think that all scientific progress can be reduced to mechanisms, machines, gear boxes ... though these things too have their own beauty ... "[12]

Marie Curie's life is a challenge to many of us who claim to be Christian. There was single-mindedness, dogged persever-

172

Science has great beauty

ance and sheer hard work that bore fruit in discovery and wonder. All these have a part in Christian experience. However, the Christian life grows out of relationship with Christ rather than from dedication to a cause.

Notes

(1) S. Quinn   *Marie Curie. A Life*   Mandarin   1996   p 82
(2) S. Quinn   p 98
(3) S. Quinn   p 115
(4) J. Senior   *Marie and Pierre Curie*   Sutton   1998   p 34
(5) & (6) S. Quinn   p 171
(7) F. Giroud   tr. L. Davis   *Marie Curie A Life*   Holmes & Meier   1986   p 94
(8) *Nobel Foundation Directory, 1999 – 2000*   p 80
(9) S. Quinn   p 198
(10) *Nobel Foundation Directory, 1999 – 2000*   p 117
(11) F. Giroud   p 274
(12) F. Giroud   p 270.

39 – Some very big ideas

ERNEST RUTHERFORD : 1871 – 1937

Ernest laughed. "That's the last potato I'll dig", he cried, dropping the spade. His mother had just brought him the news that he had been awarded a Great Exhibition Scholarship to study in England. At the time, his father was growing and milling flax, but the land had to produce food for the family too. The young New Zealander already had a double first degree in Mathematics and Physical Science from Canterbury College, in Christchurch. He followed this with research, which so impressed his professors that they had recommended him for the scholarship. The award encouraged him to apply to work under J.J. Thomson at the Cavendish Laboratory in Cambridge and, having borrowed money for his passage to England, he arrived in Cambridge in the autumn of 1895.

Rutherford's arrival on the scientific stage was singularly well timed, and his career proved remarkably fruitful from beginning to end. Only three months after he arrived in England, the discovery of X-rays astounded the scientific community and "J.J." asked his research student to join him in a study of how this new radiation affected the passage of electricity through gases. Then came the further drama of Becquerel's discovery of radioactivity, and Rutherford took on board the study of yet another type of radiation, which launched him into his life's work.

While at university in Christchurch, Ernest had fallen in love

Some very big ideas

with his landlady's daughter, Mary Newton. His letters to her give a vivid account of these years. He was ambitious and confident. On October 30, 1896, he wrote, "I am working very hard in the Lab, and have got on what seems to me a very promising line – very original needless to say. I have some very big ideas which I hope to try and these, if successful, would be the making of me."[1] Less than three years after arriving in England, he was appointed Professor of Physics at McGill University in Montreal. J.J.'s testimonial must have helped. "I have never had a student with more enthusiasm or ability for original research than Mr Rutherford."[2] A millionaire benefactor had provided a splendid physics building and facilities for McGill, and Rutherford made the most of them. He worked, with Frederick Soddy, on experiments that enabled them to suggest a theory of radioactivity. One of these experiments although using crude improvised apparatus has been described as "one of the most amazing in the whole history of science."[3]

In May 1900, nearly five years after leaving New Zealand, Ernest returned, to marry his patient fiancée. She had visited England over the summer of 1897, but for the rest of those years their love had been kept alive by letters. In the autumn, they travelled back to Canada to set up their first home together. The following spring, their only child, Eileen, was born.

In the first edition of his book on radioactivity, published in 1904, Rutherford recognised that, theoretically, vast amounts of energy could be released from small amounts of matter. He is recorded as having said that those who suggested that this might happen on an industrial scale were talking moonshine. Even so, he expressed the hope " ... that a way to release atomic energy would not be found until Man was living at peace with his neighbours."[4] One can imagine his feelings had he lived until 1945.

Rutherford had always hoped to return to England; a hope

CHAPTER THIRTY-NINE

fulfilled in 1907, when he moved to Manchester University as Professor of Physics. For the work done before this move he was awarded the 1908 Nobel Prize, for *chemistry,* much to his amusement. In Manchester, some work he suggested to one of his research students gave a totally unexpected experimental result. Rutherford described it as " ... quite the most incredible event that has ever happened to me in my life."[5] Perhaps, as with other events in these pages, you will be able to read more elsewhere,[6] for this was the clue which led him to his Nuclear Theory of the Atom, surely his greatest discovery.

In 1919, Rutherford returned to Cambridge as Professor of Experimental Physics, and Director of the Cavendish Laboratory. Part of his genius lay in attracting gifted students to his laboratory wherever he worked. Most of them adored him, although he could exhibit black moods, usually short-lived. "Perhaps the greatest single factor in Rutherford's success as a leader was his own obvious and enormous delight in experimentation"[7]. Chadwick, who was later to discover the neutron, recorded that, "To work with him was a continual joy and wonder."[8] In the words of another student, the desire to please him owed something to the fact that "Rutherford pleased was a vastly more comfortable figure to have in the laboratory than Rutherford displeased. ... Newcomers soon learnt that the sight of Rutherford singing lustily 'Onward Christian Soldiers' (recognisable chiefly by the words) as he walked round the corridors was an indication that all was going well."[9]

In later years the Rutherfords enjoyed retreating from Cambridge to a little cottage in Wiltshire. There, Lady Rutherford helped her husband to lay aside the ever-accumulating weight of honours with remarks like, "Ern, you're dribbling!" or "Ern, you've dropped marmalade down your jacket."[10] Her much-loved Ern never retired. In October 1937 he was taken ill, operated upon, and died a few days later.

Some very big ideas

During his last years, as the Nazi cancer spread through Germany, Rutherford had worked hard to help academics who lost their posts. When untimely death cut short his efforts on their behalf, there were already some 1600 professors and residential students who had been displaced.

It seems Rutherford never spoke or wrote about religion but the hymn singing would have been familiar in childhood. "Sunday was a day of complete manual rest, when the mother would regularly assemble the large family for religious exercises, she playing the piano and the father the violin, ... ". His governess records that, "Ernest was a lively boy and I loved to see him and the other children at their mother's knee at prayers."[11] Mrs Rutherford's prayerful concern for her children was lifelong. She wrote to Ernest in 1917: "That you may rise to greater heights of fame and live near to God like Lord Kelvin is my earnest wish and prayer ... "[12] She would have been encouraged by his response to a physicist who referred to the nuclear reactions on which he was working as "my reactions". Rutherford retorted, "Are you God that you call them '*my* reactions'?"[13]

CHAPTER THIRTY-NINE

Notes

(1) A. Eve *Rutherford* Cambridge University Press 1939 p 39
(2) ed. C. Gillispie *Dictionary of Scientific Biography* Charles Scribner's Sons NY 1981 Vol. 12 p 27
(3) N. Feather in, ed. M. Bunge and W. Shea *Rutherford and Physics at the Turn of the Century* Dawson 1979 p 74
(4) A. Eve p 253
(5) H. Robinson in *Rutherford (By Those Who Knew Him)* The Physical Society Memorial Lecture 1942 p 2
(6) See, for example, J. Gribbin *Almost Everyone's Guide to Science* Weidenfeld & Nicolson 1998 or B.Bryson *A Short History of Everything* Doubleday 2003
(7) H. Robinson p 15
(8) M. Oliphant *Rutherford – Recollections of the Cambridge Days* Elsevier 1972 p 69
(9) H. Robinson p 16
(10) M. Oliphant p 128
(11) E. Marsden in *Rutherford (By Those Who Knew Him)* Memorial Lecture 1949 p 46/47
(12) A. Eve p 261
(13) M. Oliphant p 145.

40 – The excitement and delight of a truly new idea

LISE MEITNER : 1878 –1968

"Once, when Lise was still very young, her grandmother warned her never to sew on the Sabbath, or the heavens would come tumbling down. Lise was doing some embroidery at the time and decided to make a test. Placing her needle on the embroidery, she stuck just the tip of it in and glanced anxiously at the sky, took a stitch, waited again, and then, satisfied that there would be no objection from above, contentedly went on with her work."[1] This incident may have been Lise's first adventure in science, testing grandma's authority, in this case groundless, against experimental evidence.

Lise was born during November 1878 – the date is uncertain – in Vienna. There, Philipp and Hedwig Meitner registered their children with the Jewish community. Lise remembered gratefully her parents' uncommon goodness and the intellectually stimulating home that they provided. At the time, education for girls stopped at age fourteen, and Lise reluctantly went on to train as a teacher. Happily, when she was twenty-one, the barriers to further education came down and after two years of determined preparation, she entered the University of Vienna in October 1901. The university was already active in research into radioactivity and Lise was fortunate in her teachers. In old age she recalled Boltzmann's

CHAPTER FORTY

lectures as "the most beautiful and stimulating that I have ever heard ... one left every lecture with the feeling that a completely new and wonderful world had been revealed."[2] Five years later she became Dr Meitner, but there was nowhere for a woman to go in science. So she made her own way.

She attended Planck's lectures in Berlin and he became her hero. It was not only physics which drew her to Planck. Music was another shared passion and she was welcome at his home, where he, with Einstein and the violinist, Joachim, sometimes made music. In 1907 she began a thirty-year collaboration with the chemist Otto Hahn, investigating radioactivity, and when work went well, the laboratories were alive with Brahms and Beethoven as they sang or whistled together. In September 1908, she formally left the Jewish community and was baptised into an evangelical Protestant congregation. One wonders if this too, was Planck's influence.

Meitner was thirty-four before she obtained her first paid position, as Planck's 'Assistentin'. Until then, she had survived on a small allowance from her father. The next year, 1913, she obtained a post in the new Kaiser Wilhelm Institute. Hahn was already occupying a similar position. The First World War slowed their work. Otto served in the army and Lise was, for a time, an X-ray nurse-technician. However, she returned to the Institute in 1917 as head of her physics section. During the post-war years she strengthened her reputation as an experimental physicist of international importance, and in 1926 she became Germany's first woman university professor of physics. She also made contacts in Sweden, which were to prove significant later. Meanwhile the research continued apace.

Then came January 30, 1933, and the installation of Hitler as Chancellor. This must surely have been the most sinister day in the twentieth century considering its ghastly sequel, at first in the lives of German Jews, scientists included, but

The excitement and delight of a truly new idea

eventually worldwide. German science was torn apart. Some collaborated enthusiastically with the Nazis; others resigned their posts in protest. Some emigrated; others stayed in post facing ever-increasing dilemmas of conscience. Hundreds were dismissed. There must be so many stories of dignified integrity in the face of this Hitler-driven disaster, which deserve to be better known.

Lise clung on until 1938. She had been dismissed from her university professorship, but this did not of itself interfere with her research at the Institute, to which she was so attached. "I built it from its very first little stone; it was, so to speak, my life's work, and it seemed so terribly hard to separate myself from it."[3] In March 1938, the annexation of Austria by the Nazis left Lise without even the nominal protection of her Austrian citizenship. A chemist in the Institute denounced her.

CHAPTER FORTY

A professor demanded she leave. Most painful of all, Otto Hahn, her colleague of thirty years, sacrificed her to save his own position in the Institute.

At Easter, Lise had one last happy holiday in the countryside with friends. She wrote in their guest book some words of Goethe's, "When troubled days come into our lives, God gives us sunshine, friendship and beauty."[4]

With the refusal of a passport, her situation became increasingly desperate, and foreign scientists schemed to get her out of the country. Eventually, a Dutch scientist, Dirk Costers, whom she had met in Sweden, travelled to Germany and brought her into Holland by a little-used railway route. She had a useless Austrian passport, two small suitcases of summer clothes and no money. Back in the Institute there were experiments in progress, unanswered letters on her desk, and her lab-coat hanging forlornly. She journeyed from Holland to Denmark and on to Sweden, where a friend took Lise into her home.

In spite of Hahn's self-protective behaviour at her expense, Meitner remained extraordinarily loyal to him. Late in 1938, he met her in Copenhagen for significant discussion of their work. In Germany, he and Strassmann – the third member of the research team – continued, in her absence, the experiments they had all been working on. Strassmann considered Lise Meitner the intellectual leader of the team, and when these experiments led to the discovery of nuclear fission, it was she who first realised the full significance of what had happened. Hahn had written to her conveying their results.

At Christmas, 1938, she showed this letter to her nephew, Otto Robert Frisch. His story of that meeting "conveys the excitement and delight of a truly new idea, the first recognition that a nucleus can split, and the first understanding of how and why it does."[5] Her notice introducing nuclear fission to the world appeared in the journal *Nature* in February 1939.

The excitement and delight of a truly new idea

However, it was Hahn who, in 1946, received the Nobel Prize. Meitner attended the Nobel ceremony, believing Hahn's prize was well deserved, but wrote to a friend afterwards, "I find it quite painful that [Hahn] in his interviews did not say one word about me, to say nothing of our thirty years of work together."[6] In spite of several nominations, she never did receive the prize herself.

Lise found life in Sweden very difficult and work almost impossible. However, when invited, in 1943, to join a group of scientists in America, working on the atomic bomb, she refused to have anything to do with it. After the war she gradually rebuilt her links with the scientific community, even in Germany, and she did not fully retire until 1960. In that year, aged 82, she moved to England and settled in Cambridge to be near her nephew, admitting that physics kept slipping away from her. She was still receiving prizes up until 1966. She died on October 27, 1968, having outlived Otto Hahn by three months. Her simple tombstone can be found in the same parish graveyard as her younger brother's, at Bramley, in Hampshire. It reads, "Lise Meitner – a physicist who never lost her humanity".

CHAPTER FORTY

Notes

(1) R. Sime *Lise Meitner A Life in Physics* University of California Press 1996 p 5
(2) R. Sime p 13
(3) R. Sime p 148
(4) R. Sime p 188
(5) R. Sime p 237
(6) R. Sime p 344.

(I feel sure that, for anyone who reads German, Meitner's letters in the archives of Churchill College, Cambridge, would prove to be well worth reading.)

41 – His search for truth was uncompromising

ALBERT EINSTEIN : 1879 – 1955

"My life is a simple thing that would interest no one. It is a known fact that I was born, and that is all that is necessary."[1]

This was Albert's comment to a fifteen-year-old high school pupil seeking copy for his school newspaper. It is not a view shared by Einstein's numerous biographers and their readers. The birth was at Ulm in Germany on March 14, 1879, and in some senses the life *was* simple. He disdained material possessions, protocol, formality and fashion – this last in general but especially by not wearing socks.

One thing is clear. A good brain is no guarantee of good relationships. Just a month before his own death, Einstein wrote to the family of a friend who had died, "What I most admired about him as a human being is the fact that he managed to live for many years not only in peace but also in lasting harmony with a woman – an undertaking in which I twice failed rather disgracefully."[2] His first marriage ended in divorce and by the above admission, the second was not all sweetness and light. Some felt that, while delighting in the company of most children, he failed his own, at least as they grew older. None the less, he made deep and lifelong friendships.

CHAPTER FORTY-ONE

In his lifetime he achieved film-star status, mobbed by photographers and journalists. Today the name of Einstein is known to millions who could name few other scientists, living or dead, and who have no clue about the reasons for his fame. If asked for an image of science, many would certainly plump for the famous halo of wild, white hair familiar from the 1930s onwards. Photos also show, almost always, at least the hint of a  mischievous grin, prelude to the uproarious laughter for which he seems to have been well known. Even in 1889, a school photo taken in Munich shows ten-year-old Albert as the only smiler. But he was not happy at school. Laughter can hide pain as well as reveal happiness.

Albert's later education was in Switzerland and aged 21, he took Swiss citizenship. He alienated teachers and professors alike by "ignoring what bored him", by "his independent and disdainful manner", and by "his casual study habits". His maths teacher at Zurich Polytechnic called him "a lazy dog" and a physics instructor said, "You're enthusiastic, but hopeless at physics".[3] Books became his teachers and fed the independent thinking which was the key to his intellectual success. His domestic life was another matter. The early years after college were financially precarious, sometimes desperately so. In 1901 he fathered a child by Mileva Maric. The little girl was presumably given for adoption, for her subsequent life remains a mystery. His appointment to the patent office in Bern in June 1902 resolved the financial problem and enabled him to marry Mileva the following January.

Readers must turn to the biographies for detail of Albert's subsequent academic appointments; his conflict with and escape from Nazi Germany; his life in America from 1933 and his involvement with the atom bomb; the birth of the Jewish State and – unknown to him – the FBI. Einstein was never an orthodox Jew, or indeed an orthodox anything, but his search for truth was uncompromising. It endeared him to kindred spirits and enraged the partisan and the bigot at both extremes. His kindness was inexhaustible. He never seemed too busy to respond to those whose need struck him as genuine.

It is staggering to realise that in the first six months of 1905 Einstein submitted to a prestigious scientific journal four papers of which any *one* would have made his name. His work at the patent office seems to have allowed time for the scribbling of formulae and equations that became a lifelong obsession. Once, when asked where his laboratory was, he held up his fountain pen. It was said of him, "Do you realise that Einstein is a scientist who needs no laboratory, no equipment, no tools of any kind? He just sits in an empty room with a pencil, and a piece of paper, and his brain, thinking."[4] One of the 1905 papers was on special relativity but it was for the work on the photoelectric effect that he was awarded the Nobel Prize. The delay of the award until 1922, after eight rejections, has been attributed to the difficulty of the awards committee in understanding relativity, and to the opposition of an influential anti-Jewish German physicist.

Another German physicist, Werner Heisenberg, has said in effect that the price of progress in science has been the inability of most people to understand what is going on. "Hear Hear!" say I as I struggle with relativity. The special theory is so called because it describes events as they appear to

CHAPTER FORTY-ONE

observers under special limited conditions. At its heart is the understanding that nothing can travel faster than the speed of light, 186,000 miles per second. As this enormous speed is approached strange things happen. Clocks slow down. Rulers shrink in length. The mass of objects increases. The theory led to the awareness that mass is a form of energy and to the famous equation $e = mc^2$. It means that to find the amount of energy available in a certain quantity of matter you multiply that quantity by the velocity of light squared, i.e. 186,000 x 186,000. The amount of energy is therefore enormous, as the explosion above Hiroshima devastatingly made clear. As to the birth of the special theory, Einstein is reported to have woken one morning "in a state of great agitation, as if, he said, 'a storm broke loose in my mind'. With it came the answers."[5]

Ten years later he was at last, "with growing excitement", nearing his goal of a *general* theory of relativity. The general theory involves concepts that most of us find even more difficult to get our heads around. It is a theory of space and time which thinks of time as a fourth dimension. When added to the three dimensions of space it forms what is called a space-time continuum. It uses equations to describe the relationship between space, time and matter. It involves the idea, bizarre to most of us, that a lump of matter in space causes space itself to curve.

The idea has been illustrated as follows. "The way to get a feel for this description of the Universe, ... is to imagine suppressing two of the four dimensions, and visualising space-time as a stretched, two-dimensional rubber sheet. The effect of the presence of a massive object in space can be pictured by imagining a bowling ball placed on the stretched sheet, where it makes a dent. Small objects moving past the heavy object (marbles rolled across the sheet) follow curved paths, because of the indentation made by the heavy object. This curving of their trajectories is the process that corresponds to the force

His search for truth was uncompromising

we call gravity."[6] This understanding of gravity dawned on Einstein in late November 1915 as, "his heart palpitating, and feeling as if something in him was about to snap, he got the answer he was after. It was like emerging from the dark into the light. He was euphoric for days afterwards, having achieved what some consider to be 'the supreme intellectual achievement of the human species'."[7]

When I reflect on the power of such soaring human thought it gives even more significance to the verse from Isaiah, " 'As the heavens are higher than the earth, so are my ways higher than your ways and my thoughts than your thoughts,' declares the Lord."[8]

Notes

(1) D. Brian *Einstein – A Life* John Wiley & Sons 1996 p 276
(2) D. Brian p 422
(3) D. Brian p 18
(4) D. Brian p 392
(5) D. Brian p 60/61
(6) J. Gribbin *Almost Everyone's Guide to Science* Weidenfeld & Nicolson 1998 p 205
(7) D. Brian p 91
(8) *Isaiah* 55: 9 and 8b.

42 – Drifting continents

ALFRED WEGENER : 1880 – 1930

On his fiftieth birthday Alfred Wegener set out from Mid Ice Station, high on the Greenland ice cap, heading for the West Coast. With him were one companion, the 22-year-old Greenlander, Rasmus Villumsen, and seventeen dogs. Wegener's body was found on May 8 of the following year. Perhaps, one day, the ice will give up Villumsen and the dogs.

"Alfred Wegener was born in Berlin on November 1, 1880, the youngest child of the evangelical preacher Dr Richard Wegener and his wife Anna."[1] It was in Berlin too, that Alfred went to school and university, but the family also owned a much-loved holiday home, now a Wegener memorial, about 60 miles to the north of the city. He began his career in astronomy, but soon changed to meteorology. "In astronomy everything has essentially been done. Only an unusual talent for mathematics together with specialised installations at observatories can lead to new discoveries; and besides astronomy offers no opportunity for physical activity."[2] For a time he worked with his older brother, Kurt, both of them fascinated by balloons and kites. In 1906, they stayed aloft for fifty-two hours to break the current ballooning record by seventeen hours.

I had for a long time associated the name of Wegener with the intriguing idea that the continents have drifted around on

Drifting continents

the surface of the globe; vast rockbergs in a semi-fluid crust. It came as a surprise to learn that he was probably more famous in his lifetime as an Arctic explorer. Alfred joined his first Greenland expedition also in 1906, but the death of the expeditions' Danish leader demonstrated the risks involved. Nevertheless, he took the bait and in 1913, during a second expedition, made a dramatic 750 mile traverse of the ice sheet with another Dane, Johan Koch. Between these events his work took him on a visit to the meteorologist, Wladimir Koppen. While there, he fell for the Koppens' young daughter, Else, and they married on his return from Greenland in 1913. His letters to her in the meantime had touched on matters other than their relationship. Noting the similarity of fit between the coastal outlines of west Africa and eastern South America, he wrote, "The fit is even better if you look at a map of the floor of the Atlantic and compare the edges of the drop-off into the ocean basin rather than the current edges of the continents. This is an idea I'll have to pursue."[3]

He must have pursued it keenly, for he first proposed his continental drift hypothesis in 1912. When war broke out, Alfred served in the army. His brother Kurt records that in 1914 Alfred was shot through the arm. About fourteen days after return to duty a bullet lodged in his neck and he was never fit for active duty again.[4] The extended sick leave enabled him to give further attention to his ideas about continental drift, and in 1915, the first edition of *The Origin of Continents and Oceans* was published.

Wegener was not the first to notice the similarity of outline shared by eastern South America and western Africa. In 1858 a book was published showing a diagram of a globe with the two continents snuggled together.[5] F.B. Taylor in America devel-oped such ideas further, a little before Wegener. However, Wegener's 1912 proposal and especially the book in 1915 produced the first logically argued theory.

CHAPTER FORTY-TWO

After the war, Wegener succeeded his father-in-law at the German Marine Observatory in Hamburg. He then, in 1924, moved to Graz as Professor of Meteorology and Geophysics and became an Austrian citizen. He took great pains to revise and update his book and produced new editions in 1922, 1924 and 1929, by which time the 94-page original had grown to 231 pages. From the very beginning, Wegener's ideas attracted some support, but the suggestion that the continents had ever moved around the globe creating mountain ranges as a sort of bow wave, was to most people, plain daft. And they said so. To one, such ideas seemed the " ... delirious ravings of people with bad cases of moving crust disease and wandering pole plague." To another, Wegener's idea was, " ... the fantasy of a geophysicist ... that would pop like a soap bubble."[6]

There is the feeling that a meteorologist meddling in geological matters was not welcome. Nevertheless, the idea provoked enough interest to hold a gathering in America in 1926 at which the pros and cons were aired. A leading British geologist, reviewing the report on this gathering noted that, although it was easy to disprove sections of Wegener's theory, "The important issue now is not so much to prove Wegener wrong as to decide whether or not continental drift has occurred, and if so how and when."[7] When I tracked down a copy of this report I was not too surprised to find that, on its title page, someone had added in pencil, "or the Nonsense Book". To the majority, the idea remained nonsense until after World War II.

In 1929, Wegener led a small preparatory expedition to Greenland, followed by a major expedition the next year, when his twenty-strong team established camps on the east and west coasts, and at Mid Ice in between. Propeller-driven motorised sleds were used for the first time, as were explosives to produce shocks by which the thickness of ice above the underlying rock could be measured. From the outset, the

Drifting continents

expedition was beset by delays. They eventually cost Wegener his life when, uncomfortably late in the season, he and others attempted to get supplies to the Mid Ice Station. After Alfred's death, his brother, Kurt, travelled to Greenland to complete the expedition as planned.[8]

By 1966, when Wegener's book was at last translated into English, Professor King, in his introduction, was able to write, "The reality of continental drift is now hardly in dispute." [9] However, geology now speaks not so much about drifting continents, but sees the globe divided into vast plates with new material being added at mid-ocean ridges as volcanic furnaces pour out new ocean bed to either side. Elsewhere, plate edges are sinking back into the earth's mantle to be recycled. Earthquakes occur where others grind against each other. It is a dramatic picture, which surely would have excited Wegener no end.

There is a striking passage in a letter Wegener wrote to one of his expedition members, which the Church would do well to heed. "You will certainly not, because of a passing ill mood, make a decision which for many years, perhaps for your whole life, would cloud the memory of our expedition. ... Whatever happens, the cause must not suffer in any way. It is our sacred trust, it binds us together, it must go on under all circumstances, even with the greatest sacrifices. That is, if you like, my expedition religion, and it has been proved. It alone guarantees an expedition without recriminations."[10] The Church's collective memory has too often been clouded by decisions made because of a passing ill mood. There is also a tendency to divide over detail, rather than to bind together so that the cause of Christ should not suffer.

CHAPTER FORTY-TWO

Notes

(1) A. Wegener, tr. J. Biram   *The Origin of Continents and Oceans*   Methuen   1966   Note by Kurt Wegener   p xxv

(2) M. Schwartzbach, tr. C. Love   *Alfred Wegener*   Science Tech Publishers   1986   p 16

(3) M. Schwartzbach   p 76

(4) A. Wegener   p xxv

(5) *Nature*   1930   126   p 841

(6) M. Schwartzbach   p 108/109

(7) *Nature*   1928   122   p 431

(8) For an extraordinary story of human endurance see:
J. Georgi, tr. F. Lyon   *Mid Ice*   Kegan Paul   1934

(9) A. Wegener   Introduction   p xxiv.

(10) M. Schwartzbach   p 162

43 – Interpreting Scripture

Francis Bacon warned us not to "unwisely mingle" our learnings from God's word and God's works. None the less, such unwise mingling has been going on for at least three hundred and fifty years and perhaps very much longer. It stems from a high regard for the Bible and a concern for its truth, but combined with insistence on interpretations of Scripture which allow no contradiction or even questioning. It has resulted in a lot of confusion, pain and division in the body of Christ, the Church.

Galileo tackled this problem head on and I doubt if anything better has been written about it in the four hundred years that have since elapsed. Copernicus published his proposal of a sun-centred universe in 1543. Galileo believed it to be true to fact and not just a hypothesis, although neither of them actually had proof. Some among the clergy had no problem with such a proposal, nor with Galileo's own observations, which prompted new thinking about the nature of the universe.

For some, however, these developments were seen as a threat to their understanding of Scripture. Verse 6 of *Psalm 19* reads, "It [the sun] rises at one end of the heavens and makes its circuit to the other". Concerning this verse, Cardinal Bellarmine wrote, " ... all expositors up to now have understood [this] by attributing motion to the sun. To affirm that the sun is really fixed in the centre of the heavens and merely turns upon itself without travelling from east to west, and that the earth is situated in the third sphere and revolves

CHAPTER FORTY-THREE

very swiftly round the sun is a very dangerous thing, not only in irritating all the theologians and scholastic philosophers, but also by injuring our holy faith and making the sacred Scripture false."[1]

Galileo could see that if the Copernican proposal came to be accepted, as it has been, such statements could only damage the Church and Scripture. In a long and remarkable letter of 1605, he wrote, "To me the surest and swiftest way to prove that the position of Copernicus is not contrary to Scripture would be to give a list of proofs that it is true and that the contrary cannot be maintained at all: thus since no two truths can contradict one another, this and the Bible must be perfectly harmonious ... "[2] In other words he accepted the Bible as true and necessarily in harmony with the proven discoveries of science. He makes this point very clearly. "As to rendering the Bible false, that is not and never will be the intention of Catholic philosophers such as I am; rather, our opinion is that the Scriptures accord perfectly with demonstrated physical truth. But let those theologians who are not astronomers guard against rendering the Scripture false by trying to interpret against it propositions which may be true and might be proved so."[3]

There was lingering resistance to the Copernican system for centuries but it is unlikely that Christians could now be found who would quote Scripture against it. However, the lesson has not been learned. The theory of organic evolution is an example. It may challenge our interpretation of Scripture but most of us, including many who are scientists, are not in a position to weigh the detailed scientific evidence. You may have heard a Christian speaker say, for example, "There is no evidence for evolution." We need to remember that the evidence which led Darwin to propose his theory in the first place is set out between the covers of *The Origin of Species*. We should be aware too, that if Darwin had never lived, the theory

Interpreting Scripture

would have come to us via Alfred Wallace. We can be certain that if there were *no* evidence, then the theory would not have gained any credence at all in the scientific community. The fact is that it commands very widespread support. I have heard it said that, "Evolution is a philosophy of life that man by himself determines truth. That's really what it is." But this is not so. Aspects of science may have spawned all manner of philosophies but Evolution, as applied to Darwin's work, is a theory about the origin of species. A Christian concern for truth demands a care for accuracy, together with the rejection of misrepresentation and sarcasm.

Some Christians refer to the Big Bang Theory as "this godless theory", but no theory is godly or godless. A theory is a proposal about how some aspect of the universe works. It will stand or fall as evidence comes to support or refute it. Some will want to say, "But the Bible says ... ", just as they did in Galileo's day concerning the Copernican theory. Whatever our feelings about a particular theory, it seems wiser to say, in the spirit of Galileo, "This theory *may* be true. It might even come to be proved so. I shall refrain from quoting Scripture against it." Some may feel this is a weak or unacceptable stance, but evidence may take a lifetime, even generations, to accumulate. Today's accepted science may look very different in a hundred years' time, but not necessarily so. What some see as chance may be something very different in the mind of God. If He can be involved in a sparrow's fall (Matthew 10: 29) I see no reason to doubt His involvement at any level He chooses in a universe He has made and sustains. The fact that some scientists are aggressively dismissive of Christianity should warn us against the dangers of a closed mind. We need to beware of showing a similar attitude towards science from a position of ignorance. Far better to be ready to give a reason for the faith we hold and to be aware that people have a right to look for evidence in our lives that faith is genuine. For Christians, the battle is not with

CHAPTER FORTY-THREE

science, but, as it always has been, with apathy, agnosticism and atheism.

The Bible is unique. It has proved extraordinarily resilient in the face of criticism. William Blake's verse on Voltaire and Rousseau is apt: "Mock on, mock on, Voltaire, Rousseau; / Mock on, mock on; 'tis all in vain! / You throw the sand against the wind, / and the wind blows it back again."[4] The Bible's authority is independent of any particular group of Christians in time or place. That has not prevented particular groups of Christians from insisting their interpretation of the Bible is exclusively correct. To quote Blake again, "Both read the Bible day and night, / But thou read'st black where I read white."[4]

How are we to respond to this predicament? The Bible is God's gift to us. He does not force it upon us. We are free to turn our backs upon it. In offering our particular interpretations we must show the same grace. We may not force a particular interpretation upon others. All who accept the authority of the Bible would probably agree that we must come to it with an open mind and asking help from the Spirit of truth (John 16: 13-15). How difficult it is though to come with a genuinely open mind. For this reason alone we must be able to say to those who differ from us, "I do believe in the unadulterated word which you hold there in your hand, but you must pardon me if, in some things, I doubt your interpretation."[5] That opens the way for dialogue and the possibility of truth prevailing. Exclusion, a refusal to listen, or a rigid insistence on a particular interpretation is often used in an attempt to keep error at bay. Sadly they are all just as likely to fetter truth which cries out to be let loose in the world. If God could send his Son into the world as a vulnerable baby and his Word as a vulnerable book, can we not take similar risks with our understanding of the truth?

It seems natural enough that we associate most closely with

Interpreting Scripture

others who share our understanding of truth. We may believe that "There can be no true fellowship between people who disagree on their understanding of Jesus. To have fellowship with one another we have to have a common faith in the incarnate, crucified and risen Christ."[6] That does not mean we cannot take account of the thinking of those from whom we differ, or that we must call for their exclusion. Fellowship has many levels. George Rawson's hymn with its striking refrain still has something to say to us: "We limit not the truth of God to our poor reach of mind. ... The Lord has yet more light and truth to break forth from his word." It is worth reminding ourselves too that, in spite of all our mistakes and differences, both petty and serious, "The word of our God shall stand for ever."[7]

Notes

(1) S. Drake   *Discoveries and Opinions of Galileo*   Doubleday Anchor   1957   p 163
(2) S. Drake   p 166
(3) S. Drake   p 168
(4) J. & M. Cohen   *The Penguin Dictionary of Quotations*   Penguin   1960   p59
(5) R. Ingrams   *England. An Anthology*   Fontana   1991   p 214   (From Trollope's *Barchester Towers*)
(6) P. Beasley-Murray in *Encounter with God*   Scripture Union   Jan-Mar 2000   p 69
(7) *Isaiah* 40: 8 KJV.

44 – An expanding universe

EDWIN HUBBLE : 1889 – 1953

"Your husband's work is beautiful – and – he has a beautiful spirit."[1] So said Albert Einstein to Grace Hubble. Hubble's work was universally admired, but not everyone agreed with Einstein about Hubble's beautiful spirit. Walter Adams, Director of the Mount Wilson Observatory, had to cope with petty squabbles among his staff: "This is not the first case in which Hubble has seriously injured himself in the opinion of scientific men by the intemperate and intolerant way in which he has expressed himself."[2] Allan Sandage, an astronomer who worked closely with Hubble in his last years, "couldn't recall which was the most frightening to him in the early days of their association – Hubble or the universe."[3]

Edwin was born on November 20, 1889 to John and Virginia in Marshfield, Missouri, USA. Both were committed to their Baptist roots and the study of the Bible. Sunday mornings saw the whole family regularly in church, and it was a sizeable family. Edwin had six brothers and sisters who survived infancy. Edwin himself sang in the choir and became well versed in the Bible. After church there was swimming, or riding the hay-wagons, or sledging according to the season, with plenty of parlour games if the weather kept them indoors.

John's ill health and changes of job involved the family in many moves, which brought Edwin eventually to Chicago

An expanding universe

University. By now he stood 6' 3" tall and was a notable athlete. At Chicago, the physicist Robert Millikan supported Edwin's application for the Rhodes Scholarship, which he was awarded in 1910. He sailed for England in September, and, at Queen's College, Oxford, chose to study law rather than mathematics or science. He came to relish England and all things English, though some considered his adopted mannerisms and vocabulary affected.

It was in Oxford that he turned his back on his parents' faith. His anxious father wrote, "I also want to impress upon you the necessity of staying with your church. Do not get out of the habit of attending. It is the greatest comfort in life, and the man who thinks he can get along without his Creator, will wake up to a sad realisation of his error, perhaps too late."[4] John died before his son returned to America. At some point, Edwin had written, " ... perhaps I haven't been to the Baptist church as much as I ought. At any rate I intend going next Sunday."[5] He was not the first or the last to confuse churchgoing with the personal relationship with God that lies at the heart of Christianity. Nobody ever inherits the faith. "God has no grandchildren". As Nicodemus discovered, entry into 'the kingdom of God', 'everlasting life', 'salvation', is by a new birth[6], freely chosen by the individual in every generation. Parents inevitably affect that choice in their children. Perhaps Edwin's parents got the balance wrong and alienated their son. Perhaps, as in *Luke* 15, loving parents produced a prodigal. Perhaps a bit of both.

So far, no mention has been made of astronomy. Edwin's grandfather sowed the seed. He owned a telescope, believing that "the most beautiful things in the universe were the planets and the stars."[7] Edwin, for his eighth birthday, begged that, instead of having a birthday party, he be allowed to stay up late, to look through grandpa's telescope. On June 23, 1899, he and his friend Sam Shelton spent the night out of doors to

CHAPTER FORTY-FOUR

watch an eclipse of the moon. Aged 14, he made the most of being bedridden after appendicitis, to read about the stars. This early enthusiasm lay dormant until, over a year after his return from England, he wrote to the astronomy professor at Chicago and obtained employment at the Yerkes Observatory there. In May 1917 he was awarded a PhD for *A Photographic Investigation of Faint Nebulae*. America had just declared war and Edwin immediately volunteered for war service. He did not reach Europe until the autumn of 1918, so that most of his service was with the army of occupation and he was spared the slaughter of the trenches. He returned to the USA in August of 1919 and joined the staff of the Mount Wilson Observatory on 3rd September.

A fellow-astronomer remembered Hubble during those early nights. "His tall vigorous figure, pipe in mouth, was clearly outlined against the sky. A brisk wind whipped his military trench coat around his body and occasionally blew sparks from his pipe into the darkness of the dome. .,. The confidence and enthusiasm which he showed on that night were typical of the way he approached all his problems."[8]

Grace Burke Leib first met Edwin on a visit to Mount Wilson in 1920. After her husband died in a mining accident the following year, they renewed their acquaintance, which led to their marriage in 1924. Their only child was stillborn. It is said that Grace never

An expanding universe

met any of Edwin's family. The Hubble home in San Marino became an informal research centre. Later, Mrs Hubble recalled, "I think it was in the 1930s when about every two weeks some of the men from Mount Wilson and CalTech [the California Institute of Technology] came to the house in the evening ... astronomers, physicists, mathematicians. They brought a blackboard from CalTech and put it up on the living room wall. In the dining room were sandwiches, beer, whiskey and soda water; they strolled in and helped themselves. Sitting around the fire, smoking pipes, they talked over various approaches to problems, questioned, compared and contrasted their points of view – someone would write equations on the blackboard and talk for a bit and discussion would follow."[9] Igor Stravinsky, Aldous and Julian Huxley, and Einstein were also visitors.

On the night of October 10, 1923, Hubble's discovery of a variable star in the Andromeda nebula enabled him to prove that this nebula lies far outside our galaxy. Until then, opinion was divided. At a debate before the National Academy of Sciences in 1920, Harlow Shapley had argued against the probability of 'island universes' beyond our own galaxy; Heber Curtis had argued the case for their existence. With his usual caution, Hubble worked for weeks to be sure of his facts before writing to Shapley in February 1924. Someone in Shapley's office when this letter arrived recalls him saying, "Here is the letter that has destroyed my universe."[10] It is a dramatic example of the force of scientific evidence. Hubble's legal training perhaps helped him to state his case and Shapley had the knowledge to realise instantly that the evidence set out in Hubble's letter pulled the rug from under his long-held opinion. In this way, science moves forward. Sometimes, the evidence is less clear-cut and it may take decades before a change is generally accepted. Sometimes, even scientists will resist strong evidence if it contravenes a lifetime's accepted theory.

CHAPTER FORTY-FOUR

Early in the twentieth century, it had been discovered that some of the nearest nebulae were not stationary, but moved at great speeds. Hubble turned his attention to these and in January 1929, produced "one of the most prominent papers in the entire history of astronomy."[11] He had discovered a direct relation between the distance of galaxies from us, and their velocity. A galaxy twice as far away as another is travelling at twice its velocity. This led to the modern understanding of an expanding universe. If expanding, then it must have been smaller in the past, and had a beginning. It was the astronomer Fred Hoyle who coined the phrase – 'The Big Bang'.

Hubble loved books and assembled a notable library. He made several visits to England, preferably in the fishing season, for he enjoyed staying with friends in Hampshire and tempting trout from the River Test. By the 1930s he had become internationally known and was much in demand as a lecturer, both in the States and in England. In the Second World War he called passionately for America to declare war in support of Europe long before Pearl Harbour. Between 1942 and 1945 he abandoned astronomy for research on all manner of missiles. He was stationed on the East Coast, far from home, although Grace eventually joined him there. In 1949 he suffered a serious heart attack, but had made a good recovery before his sudden death in September four years later.

There was no memorial service. There is no known grave. When Hubble died it was said, "He will go down to history as one of the most distinguished astronomical observers of all time."[12] To most people who know the name at all, 'Hubble' is a telescope. It is said that it could "home in on a small battery-powered flashlight held by an astronomer standing on the surface of the moon."[13] The idea of a telescope of such power carrying his name around the globe would surely have delighted Hubble. Launched into orbit in 1990, its early failures were overcome and it has been feeding data and stunning images

An expanding universe

back to Earth. Some of these pictures are now widely available. Many of them are extraordinarily beautiful. The psalmist, with only his eyes to sweep the night sky, sang out: "The heavens declare the glory of God; the skies proclaim the work of his hands."[14] Can we stay silent?

Notes

(1) G. Christianson   *Edwin Hubble: Mariner of the Nebulae*   Institute of Physics Publishing   1997   p 211
(2) G. Christianson   p 233
(3) G. Christianson   p 343
(4) G. Christianson   p 79
(5) G. Christianson   p 68
(6) *John* 3
(7) G. Christianson   p 4
(8) A. Sharov and I. Novikov, tr. V. Kisin   *Edwin Hubble: Discoverer of the Big Bang Universe*   Cambridge University Press   1993   p 14
(9) N. Hetherington   *Hubble's Cosmology*   Pachart   1996   Preface, p xvi
(10) A. Sharov and I. Novikov   p 30
(11) A. Sharov and I. Novikov   p 60
(12) A. Sharov and I. Novikov   p 120
(13) G. Christianson   p 364
(14) *Psalm* 19:1.

45 – Excited and thrilled by new facts about crystals

KATHLEEN LONSDALE : 1903 – 1971

"Perhaps for my sake, it was as well that there was no testimony against a high birth rate in those days."[1] So wrote Kathleen, looking back to her arrival in this world as the tenth child of Harry and Jessie Yardley. Harry spent a long time living away from his family, at first as a serving soldier, and, from the time Kathleen was five, by choice. He died when she was twenty, but she felt she probably owed to him her scientific turn of mind.

When Kathleen was born on January 28, 1903, Harry was postmaster at Newbridge, south of Dublin. He was a heavy drinker and smoker and had no patience with his wife's religion. Jessie was a Christian, a Baptist by persuasion, and very committed to her faith and in her wish for her children to follow her. When Harry left the family, Jessie moved to Seven Kings in Essex with five-year-old Kathleen, her three older sisters and the only two brothers to survive infancy. Kathleen proved to be a bright child, winning scholarships to Ilford County High School for Girls and then on to Bedford College for Women, in London. She had an unusual memory and could apparently listen to a thirty-minute sermon and write it out, word for word, when she got home from church.

When the London University honours list for physics

Excited and thrilled by new facts about crystals

appeared in 1922, Kathleen's name was at the top. One of the examiners, W.H. Bragg, was so impressed that he invited Kathleen, still only nineteen, to join his research team at University College London. The following year he moved to the Royal Institution, taking Kathleen with him. She wrote of Bragg, "He inspired me with his own love of pure science and with his enthusiastic spirit of enquiry ... "[2] At the Royal Institution she worked in the room Faraday had used, and read many of his notebooks, surely another source of inspiration. Bragg was making a scientific study of crystals – crystallography – and Kathleen became hooked. Near the end of her life she wrote, "I have loved my work. I am now, even after 45 years of scientific research, excited and thrilled by new facts about crystals and their properties."[3] A major part of her work was X-ray diffraction. This involves passing a narrow beam of X-rays through a crystal. The rays are scattered by the atoms making up the crystal and leave a distinctive pattern when recorded on a photographic plate. With painstaking measurements and much skill in interpretation, the arrangement of atoms and molecules in the crystal may then be worked out. Kathleen developed a real flair for this work.

In 1927 she married Thomas Lonsdale, who had been a fellow-student at University College London. Right from the beginning he encouraged Kathleen to continue with her scientific work. After a few years in Leeds, they moved back to London and Kathleen was invited back to the Royal Institution. In 1945 she and Marjory Stephenson became the first women to be elected to the Royal Society. It had been founded in 1660! In 1949 she became Professor of Chemistry and Head of the Department of Crystallography at University College. Prizes and honorary degrees followed. In 1960/61 she was vice-president of the Royal Society and in 1967 she became the first woman president of the British Association for Science. These were all scientific honours. Public recognition had come in

CHAPTER FORTY-FIVE

1956 when she was made a Dame Commander of the British Empire.

By the time war came there were three children aged 9, 8 and 5 whose care, with Thomas's support, she combined with crystallography. By now they had both become Quakers and Kathleen was passionately anti-war. Those not called up for the armed services were required to register for employment connected with the war effort or civil defence work. There was no conscientious objection clause and Kathleen, on principle, refused to register. Her refusal was eventually noticed and she was fined £2.00. She refused to pay. The magistrates felt they had no option but to commit her to Holloway Prison for a month. "Do the police come for me," she wondered, "or do I just have to go to prison by myself?"[4]

Prison routine still left time for about seven hours of scientific work each day and the whole experience sowed the seeds of another lifelong passion: prison reform, to go alongside her family, crystallography and peace. "The little Irish thief, my very delightful next-door neighbour, told me truly, 'Ye'll know more when ye go out than ye did when ye came in'."[5]

Perhaps it was the same neighbour who advised her not to leave anything lying about; "There are thieves, dearie, even in here!"[6] She travelled widely. On a scientific trip to Russia in 1951 she asked to visit a prison. Upon leaving she noticed her interpreter grinning. She asked why. "The prison governor wants to know how it is such a nice lady knows so much about prisons."[7] There were problems with early visits to the USA. An embassy official explained the long delays in granting a visa: "You've been to the three most difficult places, Russia, China and gaol."[8]

Kathleen was short of stature and crowned by a fuzzy halo of hair, usually cut for her by Thomas. She had a great sense of fun. One senses that this was at the root of some of her economies, just as much as not wanting to spend unnecessarily. For her Buckingham Palace investiture it is recorded that she made her hat for the occasion out of a piece of lace, some coloured cardboard and ninepence worth of ribbons. Almost as economical was her "new dress" for her first honorary degree – a strip of beautiful material pinned inside her gown! For her last few years she and Thomas lived in Bexhill, a long journey from work, but worth it, she thought, for a more peaceful existence and her enjoyment of the countryside she travelled through. In December 1970 she was admitted to hospital and died of cancer on the following April first.

Some of her students apparently found her academic lectures incomprehensible, but she was a popular general speaker and writer on faith, science, peace, and prison reform. Much of this material survives and is easily readable and still relevant. For example:

> "Our intellect may be perplexed and baffled by the problems of evil, of disease, of natural disaster, of undeserved suffering, and yet, in spite of this, we may have a sense of God's love that changes our belief in God from a hypothesis to a conviction ... "[9]

Referring to the fact that she did not always get answers to the questions that puzzled her, she said, "But that doesn't worry me now, because I have learned, as a scientist, how much I don't understand. I have learned too that when a scientist encounters two apparently irreconcilable ideas, these are the stepping stones to new knowledge."[10]

CHAPTER FORTY-FIVE

" ... thinking for myself brought me closer to Jesus, for he had the simplicity of approach that I wanted. He didn't just talk about God, he talked with God, and he taught his friends to do the same."[11]

Notes

(1) D. Hodgkin in *Biographical Memoirs of Fellows of the Royal Society* 1975 Vol. 21 p 447
(2) D. Hodgkin p 449
(3) K. Lonsdale *The Christian Life – Lived Experimentally* An anthology selected by J. Hough Friends Home Service Committee 1976 p 14
(4) D. Hodgkin p 453
(5) K. Lonsdale p 11
(6) D. Hodgkin p 453
(7) G. Kass-Simon & P. Farnes *Women in Science* Indiana University Press 1990 p 357
(8) D. Hodgkin p 471
(9) K. Lonsdale *I Believe* Cambridge University Press 1964 p 53 (18th Eddington Memorial Lecture, 6/11/64)
(10) K. Lonsdale *The Christian Life* p 31
(11) K. Lonsdale *The Christian Life* p 20.

46 – DNA : It's so beautiful, you see, so beautiful

FRANCIS CRICK : 1916 – 2004 AND JAMES WATSON : b. 1928

We have some odd ways of saying things but we all know what is meant when someone says of a new baby, "She's got her mother's nose." Discovering *how* she got her mother's nose has taken a long time. Since Adam lay with Eve and she gave birth to Cain, the human race has known what to do to start a baby. By comparison, discovery of the egg cell and

CHAPTER FORTY-SIX

spermatozoon, and their part in the process, is recent: the microscope first revealed spermatozoa in 1677. How the egg and sperm carried the parents' features to the offspring remained a secret for much longer.

This did not prevent plant and animal breeders improving their stock by selection, and this had been going on for centuries before Mendel's work came to light in 1900. With his discovery it became clear that reproduction involves the passing on of some kind of unit which can keep its identity from one generation to the next. Johannsen named these units 'genes' in 1909. They make up the strand-like structures called chromosomes within the nucleus of every living cell. The discovery of the structure of DNA, in the second half of the twentieth century, led to the deciphering of the genetic code. Since 1988, the Human Genome Project has been mapping the position of all the human genes on the chromosomes and sequencing their DNA. The project was completed early in 2001 and provides the clearest picture yet as to which genes do what to determine the contours of the new baby's nose, and very much else besides.

It was the realisation that the nature of the gene was one of the century's big unanswered problems that caught the interest of the Englishman, Francis Crick and the American, James Watson. Each had read the book, *What is Life?* by the Austrian physicist, Erwin Schrödinger, and this had pointed the way. However, in 1951, Crick, aged 35 was only 2 years into his first real research, for his PhD. Behind him were a less-than-brilliant physics degree, war service with the Admiralty, and the wreckage of some uninspiring research. "By good fortune a land mine had blown up the apparatus I had so laboriously constructed at University College, London, so after the war I was not obliged to go back to measuring the viscosity of water."[1] In 1949 he joined the Cambridge Medical Research Council Unit of the Cavendish Laboratory under Max Perutz, researching the connection between the structure of protein molecules and how they worked.

When 23-year-old James Watson arrived in Cambridge in the autumn of 1951 he already had his PhD from the University of Chicago. He had been enduring a frustrating year of research in Copenhagen, but had visited Naples for a scientific meeting and met Maurice Wilkins. Wilkins was already trying to work out the structure of DNA, and displayed some of his X-ray diffraction photographs of the DNA molecule. James realised at once that this was the way forward, and it was Wilkins who recommended that he try Cambridge. On arrival, he recalls that he " ... immediately discovered the fun of talking to Francis Crick. Finding someone in Max's lab. who knew that DNA was more important than proteins was real luck."[2] The key things to keep in mind here are that the genes consist of DNA, not protein as most researchers thought at the time; that DNA can be crystallised and that, therefore, its structure can be investigated by X-ray diffraction.

So, we now have in Cambridge two men with their minds set on working out how the DNA molecule is put together, but, in many respects, neither of them well-qualified for the task. Watson even hoped " ... that the gene might be solved without my learning any chemistry."[3] Further, although alerted by Wilkins, he had no experience in X-ray diffraction. Crick was turned 30 before he left physics and embraced biology as a means to his goal. And yet, on March 7, 1953, they unveiled a model showing how they thought the molecule was constructed. It had such a natural elegance that colleagues invited to see it were quickly persuaded that the two had got it right. James reports that " ... we had lunch, telling each other that a structure that pretty just had to exist."[4]

They wrote an article for *Nature,* April 25, 1953, which totally conceals their personal excitement. "We wish to suggest a structure for the salt of Deoxyribose Nucleic Acid. This

CHAPTER FORTY-SIX

structure has novel features which are of considerable biological interest."[5] The page on which this article appears must be among the most-thumbed of any volume of *Nature* on the shelves of any science library, but, from the content, you could be forgiven for not realising it describes one of the most important discoveries of the twentieth century. They and Wilkins were awarded a Nobel Prize in 1962.

The way science sometimes works is wonderfully illustrated in this story. There are the seeds of an idea; for Crick, a lecture by Linus Pauling – perhaps the greatest chemist of the century – and for both men, Schrödinger's book. There is the focus on a particular unsolved problem: the nature of the gene. There is the unplanned meeting: Watson with Wilkins in Naples, and then with Crick himself in Cambridge. There is the work of others: Maurice Wilkins and Rosalind Franklin (her X-ray photographs were a crucial key, and some thought the Cambridge pair's use of them controversial), – Linus Pauling again and specifically his book *The Nature of the Chemical Bond*. Pauling himself came very close to forestalling Crick and Watson. There is the trial and error – Franklin laughed their early model to scorn; the perseverance, the mounting excitement.

When Watson thought he had found the final piece of the jigsaw of 'atoms' making up their model he showed it to a colleague who had objected to an earlier version and asked if he had any objection to this one. "When he said no, my morale skyrocketed ... it seemed almost incredible that the DNA structure was solved, that the answer was so incredibly exciting ... "[6] The invigorating effect of such discovery is perfectly caught in Watson's comment, "The following morning I felt marvellously alive when I awoke."[7] Some have described their discovery of faith in similar words. Charles Wesley is an example: "I woke, the dungeon flamed with light; my chains fell off, my heart was free."

DNA: It's so beautiful, you see, so beautiful

This brief account says more about the discovery of the structure of DNA than about the discoverers. In this I take comfort from Crick himself who wrote: "Rather than believe that Watson and Crick made the DNA structure, I would stress that the structure made Watson and Crick. After all, I was almost totally unknown at the time, and Watson was regarded, in most circles, as too bright to be really sound. But what I think is overlooked in such arguments is the intrinsic beauty of the DNA double helix. It is the molecule that has style, quite as much as the scientists."[8] I can only encourage you to explore further in what has been written on the subject to discover why the scientists got so excited about this exquisite functional design at the hub of life. Watson, speaking about the structure they discovered, said, "It's so beautiful, you see, so beautiful."[9]

Notes

(1) F. Crick *What Mad Pursuit* Penguin 1990 p 14
(2) J. Watson *The Double Helix* Weidenfeld & Nicolson 1968 p 48
(3) J. Watson p 21
(4) J. Watson p 205
(5) *Nature* Vol. 171 Jan – June 1953 p 737
(6) J. Watson p 194
(7) J. Watson p 199
(8) F. Crick p 76
(9) F. Crick p 78.

Other books:
P. Strathern *Crick, Watson and DNA* Arrow 1997
D. Newton *James Watson and Francis Crick. Discovery of the Double Helix and Beyond* Facts on File 1992
H. Judson *The Eighth Day of Creation* Penguin 1995.

47 – Magic, mathematical ingenuity ... physical insight

RICHARD FEYNMAN : 1918 – 1988

Richard Feynman was no ordinary physicist. Others also have won the Nobel Prize, but never was there another who was a bongo player, an artist, a dancer, a repairer of radios, an expert safebreaker, and a decipherer of Mayan hieroglyphics to boot. He was also much loved and very funny. I made the painful mistake of trying to read *Surely You're Joking Mr Feynman* in the sacred silence of the Bodleian Library. His mother explained that, "When Richard was little, he couldn't decide whether to be a comedian or a scientist, so he combined the two options."[1]

Richard was born on May 11, 1918, in Manhattan and brought up in Far Rockaway, New York. From his mother he learned laughter and human compassion. His sister, Joan, recalls " ... wonderful memories of evenings at the supper table when Richard was home from college and he and mother would get going. My father and I would laugh so hard that our stomachs hurt and we would beg for mercy, but they wouldn't stop until I had fallen off my chair and was literally rolling on the floor."[2] From his father he learned to question nature, rather than to label it.

When Feynman graduated from the Massachusetts Institute of Technology in 1939, he moved to Princeton where he was supervised by the physicist, John Wheeler. In 1942 he married

Magic, mathematical ingenuity ... physical insight

his childhood sweetheart, Arlene Greenbaum. Tragically she was already seriously ill. In 1943 Richard left Princeton to work at Los Alamos on the atomic bomb project. With Arlene in hospital there were regular weekend journeys to visit and in their correspondence they managed to gain a lot of fun at the censor's expense. Arlene died in 1945.

The post-war years saw several brilliant contributions to physics by Feynman. In the 1940s he produced a version of the theory of Quantum Electrodynamics, for which he eventually shared the Nobel Prize, in 1965. Quantum Electrodynamics has been said to explain everything that is not explained by gravity. Essentially, it includes electromagnetism, one of the four fundamental forces found in nature. In ten papers between 1953 and 1958, Feynman explained how liquid helium behaves as a 'superfluid', flowing without friction. David Pines described this work as, " ... that blend of magic, mathematical ingenuity and sophistication, and physical insight that is almost uniquely Feynman's."[3]

He also contributed to the understanding of other fundamental forces; the 'weak force' involved, for example, in some types of radioactive decay, and the 'strong force' holding together the protons and neutrons within an atomic nucleus. It seems that Feynman had a very visual way of understanding things and his 'diagrams' have also helped others. James Bjorken was among them. "When Feynman Diagrams arrived, it was the sun breaking through the clouds, complete with rainbow and pot of gold. Brilliant! Physical and profound! It was instant conversion to discipleship."[4]

In the early 1950s, still missing Arlene, he married again, but the marriage was a disaster and ended in divorce in 1956. Two years later, during a visit to Switzerland, he met a Yorkshirewoman, Gweneth Howarth, by Lake Geneva. It was a meeting that led to their marriage in 1960 and nearly thirty years together. During 1961 to 1963 his research was in the

CHAPTER FORTY-SEVEN

doldrums, but he turned to it again with fresh enthusiasm after James Watson sent him the typescript of his book, *The Double Helix*. He wrote to Watson, " ... you are describing how science is done. I know, for I have had the same frightening and beautiful experience."[5] In 1967 he was offered his first honorary degree, but he turned it down, and all other similar offers.

Feynman paid no allegiance to any religion, but he has much to teach us about the sheer enjoyment of this world. All his adult life he did physics for fun. He never knew how long his working week was because he could never distinguish between working and playing. He marvelled at the beauty of the world in its fine structure. Let him speak for himself: "It's an appreciation of the mathematical beauty of nature, of how she works inside; a realisation that the phenomena we see result from the complexity of the inner workings between atoms; a feeling of how dramatic and wonderful it is. It's a feeling of awe ... "[6] "I'm always looking, like a child, for the wonders I know I'm going to find – may be not every time, but every once in a while."[7] "The same thrill, the same awe and mystery, comes again and again when we look at any question deeply enough."[8] " ... the imagination of nature is far, far greater than the imagination of man."[9] Christians need alter only one word in that last sentence. Feynman once said, "The God of the church isn't big enough"[10], but he dismissed religion as wishful thinking, thereby denying himself the opportunity of getting to know Him. And did he never read *Psalms*?

In 1978 he suffered the onset of cancer and was always grateful for the skill of his surgeon, which gave him ten extra years. He had long been well known in scientific circles when he was appointed to the Commission investigating the 1986 Challenger Space Shuttle Disaster. His simple experiment, live on television, showing how the seal on the O-rings might have

Magic, mathematical ingenuity ... physical insight

failed at low temperature, was a dramatic demonstration of the effectiveness of a scientific approach. It added public and international acclaim to that which he had long enjoyed in the world of science. He gives a vivid account of his involvement in *Mr Feynman goes to Washington.*[11]

It may come as a surprise to some to find Feynman putting great emphasis on ignorance and doubt. "We have found it of paramount importance that in order to progress we must recognise our ignorance and leave room for doubt. Scientific knowledge is a body of statements of varying degrees of certainty – some most unsure, some nearly sure, but none <u>absolutely</u> certain."[12]

Richard Feynman was active almost to the end, but the cancer finally overtook him on February 15, 1988. John and Mary Gribbin pay this tribute: " ... by teaching people to think, insisting on scrupulous honesty and integrity, never fooling yourself and always rejecting any theory, no matter how cherished, if it disagrees with experiment, and, above all, inspiring an awe and appreciation of nature and a love of science, Feynman made a mark on science, which will last whatever happens to science itself."[13]

CHAPTER FORTY-SEVEN

Notes

(1) J. Gribbin and M. Gribbin *Richard Feynman: A Life in Science* Penguin 1998 Prologue p xvii
(2) J. Gribbin and M. Gribbin p 7
(3) J. Gribbin and M. Gribbin p 157
(4) J. Gribbin and M. Gribbin p 197
(5) J. Gribbin and M. Gribbin p 185
(6) J. Mehra *The Beat of a Different Drum* Clarendon 1996 p 580
(7) R. Feynman, as told to R. Leighton *What Do You Care What Other People Think?* Grafton 1992 p 16
(8) R. Feynman p 243
(9) R. Feynman p 242
(10) R. Feynman *The Meaning of it All* Penguin 1998 p 40
(11) See chapter in *What Do You Care ... ?* (7) above.
(12) R. Feynman *What Do You Care ... ?* p 245
(13) J. Gribbin and M. Gribbin p 279.

48 – Remotest areas of the Amazon rainforest

GHILLEAN PRANCE : b. 1937

"It appears, by the dung that they drop upon the turf, that beetles are no inconsiderable part of their food."[1] When Ghillean, as a child, read this sentence about hedgehogs in Gilbert White's *Natural History*, he went out and found some droppings and searched in them for remains of beetles. Perhaps in that moment, a scientist was born. He was among the many thousands who have read that delightful book, and been inspired to take a closer look at the world around them. However, he did not become a zoologist. It seems that the influence of four wonderful aunts, all in their seventies and experts on plants, swung the pendulum of his interest towards botany.

The boy born on July 13, 1937, at Brandeston in Suffolk was christened Ghillean but became popularly known as Iain. The family spent the war years on the Isle of Skye but then moved south again, this time to Toddington in Gloucestershire. Sadly, Iain's father died soon after, leaving his mother a widow for the second time. From the age of thirteen, Iain was educated at Malvern College. His housemaster and biology teacher there was Bill Wilson, a man with a passion for field botany. His enthusiasm was contagious and Iain went on to read botany at Keble College, Oxford.

CHAPTER FORTY-EIGHT

Oxford had a surprise in store. Iain attended the University Church, although his religion at the time was little more than a formality. There came a day, however, when he faced a challenge to make a personal response. The occasion was a sermon from Canon John Collins. The Canon invited anyone who was fully persuaded of the truth of Christianity to 'Come forward'. Iain accepted the Canon's invitation, and recalls that event now as "A walk into the arms of my Saviour."[2] It proved a very significant turning point. Even so, it was not easy for him to decide to serve that summer on a Children's Mission instead of botanising. It proved to be a good decision, for it was there that he met Anne Hay, who, in 1961, became his wife.

Anne became the breadwinner while her new husband worked for his PhD, but, before they were married, Iain had led a botanical expedition to Turkey. It set the agenda for much of his professional life. Prance was already hooked on taxonomy, that part of botany which deals with the classification and naming of plants. It led to his joining an expedition to Surinam, not long after the birth of their first child, Rachel, in 1963. It was Anne's first taste of being a 'rainforest widow'.

The expedition was run by the New York Botanical Garden and Iain was invited to join the staff there immediately on return. Anne was left to pack up house and bring young Rachel to join him in America. After only seven months, he was off again, to Brazil this time. The four months that he was away were the worst of Anne's life, and she vowed to accompany him thereafter. She kept to her vow and developed a significant role in the expeditions, although, with young children to care for, life was hard. A second daughter, Susan, was born in 1965. When Iain next took to the expedition trail, she remained behind at the base in Manaus, where the life was just a little more bearable than in far-away New York.

During these years, Prance led expeditions from Manaus to

Remotest areas of the Amazon rainforest

some of the remotest areas of the Amazon rainforest. One involved nearly two and a half thousand miles of travel by boat and boot. He introduced inflatables to cut journey times and believes himself to be the only professor of botany with a certificate in outboard motor maintenance. The expeditions collected tens of thousands of carefully recorded plant specimens, many new to science, and with quite unknown potential. 95% of the world's plants have still to be tested for their medical properties. He greatly respected the Amerindian tribes they encountered and widened his research to include their botanical knowledge. A visit to the Trans Amazonian Highway left him appalled at the ecological cost and turned him into a campaigner for ecologically sensitive policies where transport, farming and development are involved.

There were adventures galore. On one occasion, fish from the river and Prance's botanical knowledge combined to save a party of 9, after the plane which had come to fetch them crashed on the primitive airstrip and left the party stranded in the jungle for several days. On another, when a virulent form of malaria struck the expedition in remote forest, he believes it was only the grace of God, in the form of local missionaries, which saw them through. A great surprise for Iain came when they camped near the homes of some freelance rubber-tappers. One of them had become a Christian through listening to broadcasts from a Christian radio station. He spotted Iain reading his Bible and was so excited to have a real Christian in front of him instead of just a voice over the air that he asked Iain to preach the next day. The expedition was delayed long enough to hold the very first live service of Christian worship in that tiny community.

One of his research projects was prompted by questions from his students which he realised he couldn't answer. How is the Royal Waterlily pollinated? It led to long hours of night-time observation with a torch while standing deep in water to find

CHAPTER FORTY-EIGHT

out just exactly what the visiting beetles were up to. His biographer writes that, for Iain, " ... the sight of a lake full of lily flowers slowly unfolding in the half-light of dusk is one of the most breathtaking he has ever seen."[3] This waterlily is one of the big attractions of Kew Gardens, and it was Kew, with an invitation to become its Director, that drew him back to England in 1988, after 24 years of service to the New York Garden.

Some of us remember with affection the penny entrance fee to Kew Gardens, but Prance had to make it more financially independent; hence the shop and the more realistic charges. Perhaps his most important scientific innovation is the Millennium Seed Bank, further reflecting his interest in conservation. Peter Crane, Iain's successor at Kew, reports that, "We hope to have 10 percent of the world's flora – 24,000 species – in the seed bank by 2010."[4] In 1994, Iain Prance was the first recipient of the International Cosmos Prize, for his "work on harmony between humankind and the environment". He is passionate about the created world and our need to care for it. "It is vital, not only for nature, but for the future health of the church itself, that we have a well-developed theology of creation and of Christian stewardship. ... the church must give a lead."[5]

Notes

(1) G. White   *The Natural History of Selborne*   Dent   1949   p 78
(2) C. Langmead   *A Passion for Plants*   Lion   1995   p 31
(3) C. Langmead   p 143
(4) *The Times* 25/8/99
(5) G. Prance   *The Earth Under Threat: A Christian Perspective*   Wild Goose   1996   p 66.

49 – Trying to discern truth

FRANCIS COLLINS:
b. 1950

The cells in your body are numbered in millions of millions. Apart from the egg or sperm cells each one of these body cells normally has forty-six chromosomes in its nucleus. They make up twenty-three matching pairs, one of each pair from your mother and the other from your father. The DNA strands that make up the chromosomes include in their structure molecules called bases. There are four kinds of these and the order in which they are arranged provides a code by which instructions are passed on to the cell. The now defunct Morse code provides an illustration. It was easy to represent any letter of the alphabet by a simple arrangement of dots and dashes, just two symbols. Out of these, any word could be built and therefore, any message passed on. DNA has *four* symbols, the bases, arranged in groups of three, and the particular sequence of these groups gives each gene its uniqueness. The gene gives the instructions that are passed on to the cell, stage by stage, to build specific proteins; and proteins are the building blocks of our bodies and the way they work. It is a bit like 'The King's Breakfast', although as the message pass through the cell there is more 'tell' and less 'ask' than in the Royal Palace.

> The King asked
> The Queen, and
> The Queen asked
> The Dairymaid:
> "Could we have some butter for
> The Royal slice of bread?"
> The Queen asked
> The Dairymaid,
> The Dairymaid
> Said: "Certainly,
> I'll go and tell
> The cow
> Now
> Before she goes to bed."[1]

The complete sequence of bases for all the genes for all the chromosomes in a human cell is what makes up the human genome. The sequence contains some three billion bases and the Human Genome Project has been working out the order in which they occur. The project began in America in 1988 under the direction of James Watson. His successor is Francis Collins who became the director of The National Center for Human Genome Research in 1993. Research Centres in other countries became involved, including The Sanger Centre near Cambridge in the UK, and a privately-funded initiative in the US. Success with a draft genome was achieved early in 2001, and an announcement made, perhaps under political pressure. It was not until April 14, 2003 that completion was finally announced. The project has been hailed as one of the great scientific ventures of our time; some would say the greatest.

Francis was born in 1950 and raised on a small farm in

CHAPTER FORTY-NINE

Staunton, Virginia. He and his brothers were educated at home by their mother until, at the age of twelve, Francis moved on to the local High School. A chemistry degree at the University of Virginia and a PhD at Yale followed. Next, he changed direction completely to enrol in the University of North Carolina's medical school. Here he became enthralled by medical genetics. Enthusiasm for research in human genetics took him back to Yale and then, in 1984, to the University of Michigan. In 1989 his research team, with Tsui and Riordan of the Hospital for Sick Children in Toronto, identified the gene for cystic fibrosis. This triumph was then followed by other successes, leading up to the 1993 appointment.

The move to medical school was not the only change of direction. His wife, Mary Lynn, had become a Christian and Francis "agreed, with considerable skepticism, to re-examine the faith he had abandoned without a great deal of thought some years before."[2] Speaking in Cambridge, England, in 1998 he briefly retraced his own spiritual journey.

"I was raised in a home where faith was not considered particularly relevant, sent to church to learn music, but instructed that it would be best to avoid the theology. I followed those instructions well and went off to college with only the dimmest idea of what saving faith in Jesus Christ was all about. What little glimmers of faith I might have possessed were quickly destroyed by the penetrating questions of my freshmen dorm colleagues who, as one will do at that phase in life, took great delight in destroying any remnants of superstition, which is what they considered faith to be. I went on to study chemistry and physics, and became quite an obnoxious atheist with whom you would not have enjoyed having lunch. This was because I too felt it was part of my mission to point out that all that really mattered could be discovered by science and everything else was irrelevant. And then, somewhat troubled by my own discovery that chemical physics didn't suit me as well as I thought, I went off to medical school to try to learn some other skills. It was also to keep my options open, being in somewhat of a disarray about how I wanted to spend my life. Only several years after that, as a medical student, watching people with terrible illnesses struggle with those challenges, I began to wonder how it was that some of the people I encountered seemed to derive such strength from something which I viewed as nothing more than superstition. I realised that though I had spent much of my life trying to discern truth, using the tools of science, I'd never really considered the evidence for the truth of faith. Fortunately through the guidance of some very patient people, who tolerated a lot of my insolent questions, I was led to read C.S,Lewis[3] and then the Bible. This led me to understand many of the concepts that had completely eluded me before, and I gave my life to Christ twenty years ago."[4]

During this same Cambridge address, Francis Collins outlined the goals and successes of the Human Genome Project. He also made it clear that there were challenges ahead for medicine, for morals and for faith. He was right. Only six years on from that talk, "The Times" of June 17, 2004, had these headlines: "Gene treatment stops frisky voles being love rats" and "Scientists await first embryo cloning licence". The article under the first begins, "A single 'love gene' that transforms Don Juans into loyal and attentive spouses has been identified by scientists, ..." The rest of the article might as well not have been written, for the opening sentence allows us to think that, "It's all in the genes", and we don't have any responsibility for our sexuality or fidelity. Genes play a very important part in making us the person we are, but they are not all-powerful. A conversion experience can transform character.

Collins is another in the long line of those whose lives give the lie to claims that science and faith are incompatible. "We must be prepared and willing to make a reasoned presentation

CHAPTER FORTY-NINE

of our faith, especially to young scientists, who have all too often concluded that a serious faith in a personal God and objective scientific truth are incompatible. We know that not to be true, and yet most young scientists have never heard those arguments."[5]

Notes

(1) A.A. Milne, "The King's Breakfast", in any edition of *When We Were Very Young*
(2) G. Liles in *MD* March 1992 p 44
(3) C.S. Lewis *Mere Christianity* E.g. Collins Fontana Books 1960
(4) F. Collins *The Human Genome Project* in *Science & Christian Belief* October 1999 p 99
(5) F. Collins p 110.

50 – A New Creation

Having read through this selection of forty or so scientists you may be thinking how your choice would differ, especially if you are from a country other than those represented; or a zoologist! It is unlikely in the extreme that a list of scientists equal in number to those featured here, but chosen by someone else, would coincide exactly, but a few would be on anyone's list. My hope is that you have enjoyed the glimpses into these lives in science and will delve deeper.

Life is many stranded. Two of the richest in my own have been drawn from the faith I placed in Jesus as a lad of fifteen years, and from my fascination with this extraordinary universe in which we find ourselves. Although my experience lies a long way from science's cutting edge, these two strands, representing faith and science, can also be followed in several of these famous lives. Are they incompatible? I do not think so. They weave inextricably through the fabric of life.

If you were unfamiliar with science, I hope that what it is has become a little clearer. The story of science through nearly three thousand years is represented in the people that you have been introduced to in these pages. You have seen their attention caught by something which most of us might have overlooked. That led them into some new line of enquiry about the universe. They made further observations or teased out a novel thought. They perhaps set up experiments with results that supported their thinking or changed its direction. A day came when they shared their discovery with other scientists. Some of these then looked for more evidence and, as a result,

CHAPTER FIFTY

ideas were accepted, modified, or rejected. In this way, by the patient accumulation of evidence, the consensus that we call 'Science' has developed.

But this is not all. Remember Dirac? "Physical Laws should have Mathematical Beauty." Curie? "Science has great beauty." Watson? "It's so beautiful, you see, so beautiful." There is an expectation that this world not only has a basis in reason, but that it also has an intrinsic beauty, from the interaction of atomic particles, through the snowflake and the solar system, to the furthest spiral galaxy. Sometimes the beauty is visible to the untrained eye but this is not the touchstone. Henri Poincaré, a famous French mathematician, described this as beauty "which a pure intelligence can grasp" rather than anything we appreciate by our senses.[1] Not being a 'pure intelligence' I miss the full impact of the beauty which lies in the elegance of a mathematical expression or in a simple law summing up a wide-ranging truth. But it gives me a kind of happiness to know it is there. It seems in keeping with the idea of Kepler "thinking God's thoughts after him".

Over the last two centuries in particular, science has been increasingly successful in discovering how the world works. In the century ahead it may probe the mysteries of time and consciousness in ways we can only guess at today. That very success has encouraged some scientists to suggest that science is the only valid route to knowledge. However, as Richard Feynman has pointed out, "Scientists take all those things that can be analysed by observation, and thus the things called science are found out. But there are some things left out, for which the method does not work. This does not mean that these things are unimportant. They are, in fact, in many ways the most important."[2]

I believe that the most important discovery any of us can make is that the same God who was "in the beginning" offers us in Christ the possibility of a new beginning. When Iain

A new creation

Prance listened to an Oxford sermon in the 1950s, he became convinced of the validity of Christ's claim upon his life. In believing, he took a step of faith, and a whole string of discoveries followed. Such discoveries are awaiting anyone who takes this step. Among them, for example, we find that, "If we confess our sins he is faithful and just and will forgive us our sins ... "[3]; that " ... the Word of God is living and active."[4]; that "God is light; in him there is no darkness at all"[5]; that "God is love."[6]. There are many more. The step of faith leads to life being wholly refocused. No wonder Paul uses the phrase "A New Creation" to describe such a life-changing event. It is interesting that those describing the discovery of new life in Christ and those describing their feelings on making a scientific discovery may do so in similar words.

As I look out on the garden today, it is brilliant with winter sunshine. Yellow jasmine and winter-flowering honeysuckle, snowdrops and crocuses enliven the scene. Whatever the scientific detail of how they came to be, I believe that the God whom I have come to know as my heavenly Father is the mind behind it all. My heart sings and the Bible lends words for my worship. "You are worthy, our Lord and God, to receive glory and honour and power, for you created all things, and by your will they were created and have their being."[7]

The Bible begins with creation. "In the beginning God created the heavens and the earth." At its heart it points to the possibility of the individual life being newly created in Christ. "So if anyone is in Christ, there is a new creation: everything old has passed away; see, everything has become new!"[8] As it ends, in *Revelation*, John is trying to find words to describe the indescribable. "Then I saw a new heaven and a new earth, for the first heaven and the first earth had passed away. ... There will be no more death or mourning or crying or pain, for the old order of things has passed away. He who was seated on the throne said, 'I am making everything new!'"[9]

CHAPTER FIFTY

Our time here is clouded with suffering, ugliness, brutality. We know that one day our earthly lives will end, and that even the Earth, for all its beauty, is not immortal. One day the mountains on the moon will be no more. The Bible closes with a vision eclipsing the material world of our senses. Science has wonderfully enriched our understanding of that world. I have found some of the images from the Hubble telescope so breathtakingly beautiful as to move me to tears. No possible description in words could have the same impact. It helps me to understand that John was given a glimpse of something beyond our imagining. In Paul's words, "No eye has seen, no ear has heard, no mind has conceived what God has prepared for those who love him."[10] Kepler celebrated God in astronomy. What a celebration awaits the people of God in his new creation.

Notes

(1) M. Bragg *On Giants' Shoulders* Sceptre 1999 p 205
(2) R. Feynman *The Meaning of it All* Penguin 1998 p 16
(3) *1 John* 1: 9
(4) *Hebrews* 4: 12
(5) *1 John* 1: 5
(6) *1 John* 4: 16
(7) *Revelation* 4: 11
(8) *2 Corinthians* 5: 17 (NSRV)
(9) *Revelation* 21: 1,4,5
(10) *1 Corinthians* 2: 9.